T0296461

CAMBRIDGE LIBRARY COLLECTION

Books of enduring scholarly value

Darwin

Two hundred years after his birth and 150 years after the publication of 'On the Origin of Species', Charles Darwin and his theories are still the focus of worldwide attention. This series offers not only works by Darwin, but also the writings of his mentors in Cambridge and elsewhere, and a survey of the impassioned scientific, philosophical and theological debates sparked by his 'dangerous idea'.

The Life and Letters of Charles Darwin

This three-volume life of Charles Darwin, published five years after his death, was edited by his son Francis, who was his father's collaborator in experiments in botany and who after his death took on the responsibility of overseeing the publication of his remaining manuscript works and letters. In the preface to the first volume, Francis Darwin explains his editorial principles: 'In choosing letters for publication I have been largely guided by the wish to illustrate my father's personal character. But his life was so essentially one of work, that a history of the man could not be written without following closely the career of the author.' Among the family history, anecdotes and reminiscences of scientific colleagues is a short autobiographical essay which Charles Darwin wrote for his children and grandchildren, rather than for publication. This account of Darwin the man has never been bettered.

Cambridge University Press has long been a pioneer in the reissuing of out-of-print titles from its own backlist, producing digital reprints of books that are still sought after by scholars and students but could not be reprinted economically using traditional technology. The Cambridge Library Collection extends this activity to a wider range of books which are still of importance to researchers and professionals, either for the source material they contain, or as landmarks in the history of their academic discipline.

Drawing from the world-renowned collections in the Cambridge University Library, and guided by the advice of experts in each subject area, Cambridge University Press is using state-of-the-art scanning machines in its own Printing House to capture the content of each book selected for inclusion. The files are processed to give a consistently clear, crisp image, and the books finished to the high quality standard for which the Press is recognised around the world. The latest print-on-demand technology ensures that the books will remain available indefinitely, and that orders for single or multiple copies can quickly be supplied.

The Cambridge Library Collection will bring back to life books of enduring scholarly value across a wide range of disciplines in the humanities and social sciences and in science and technology.

The Life and Letters of Charles Darwin

Including an Autobiographical Chapter

VOLUME 2

CHARLES DARWIN
EDITED BY FRANCIS DARWIN

CAMBRIDGE
UNIVERSITY PRESS

CAMBRIDGE UNIVERSITY PRESS

Cambridge New York Melbourne Madrid Cape Town Singapore São Paolo Delhi

Published in the United States of America by Cambridge University Press, New York

www.cambridge.org
Information on this title: www.cambridge.org/9781108003452

This edition first published 1887
This digitally printed version 2009

ISBN 978-1-108-00345-2

FROM A PHOTOGRAPH (1874?) BY CAPTAIN L. DARWIN, R.E. ENGRAVED FOR THE
'CENTURY MAGAZINE,' JANUARY 1883.

FROM A NOTE-BOOK OF 1837.

led to comprehend true affinities. My theory would give zest to recent & Fossil Comparative Anatomy : it would lead to study of instincts, heredity, & mind heredity, whole metaphysics, it would lead to closest examination of hybridity & generation, causes of change in order to know what we have come from & to what we tend, to what circumstances favour crossing & what prevents it, this and direct examination of direct passages of structure in species, might lead to laws of change, which would then be main object of study, to guide our speculations.

228

THE

LIFE AND LETTERS

OF

CHARLES DARWIN,

INCLUDING

AN AUTOBIOGRAPHICAL CHAPTER.

EDITED BY HIS SON,

FRANCIS DARWIN.

IN THREE VOLUMES:—VOL. II.

LONDON:

JOHN MURRAY, ALBEMARLE STREET.

1887.

TABLE OF CONTENTS.

VOLUME II.

ILLUSTRATIONS.

ERRATA.

LIFE AND LETTERS

OF

CHARLES DARWIN.

———•◦•———

CHAPTER I.

THE FOUNDATIONS OF THE 'ORIGIN OF SPECIES.'

[IN the first volume, p. 82, the growth of the 'Origin of Species' has been briefly described in my father's words. The letters given in the present and following chapters will illustrate and amplify the history thus sketched out.

It is clear that, in the early part of the voyage of the *Beagle* he did not feel it inconsistent with his views to express himself in thoroughly orthodox language as to the genesis of new species. Thus in 1834 he wrote * at Valparaiso: "I have already found beds of recent shells yet retaining their colour at an elevation of 1300 feet, and beneath the level country is strewn with them. It seems not a very improbable conjecture that the want of animals may be owing to none having been created since this country was raised from the sea."

This passage does not occur in the published 'Journal,' the last proof of which was finished in 1837 ; and this fact harmonizes with the change we know to have been proceeding in his views. But in the published 'Journal' we find passages which show a point of view more in accordance with orthodox

* MS. Journals, p. 468.

theological natural history than with his later views.   Thus, in speaking of the birds Synallaxis and Scytalopus (1st edit. p. 353 ; 2nd edit. p. 289), he says : "When finding, as in this case, any animal which seems to play so insignificant a part in the great scheme of nature, one is apt to wonder why a distinct species should have been created."

A comparison of the two editions of the 'Journal' is instructive, as giving some idea of the development of his views on evolution.   It does not give us a true index of the mass of conjecture which was taking shape in his mind, but it shows us that he felt sure enough of the truth of his belief to allow a stronger tinge of evolution to appear in the second edition.   He has mentioned in the Autobiography (p. 83), that it was not until he read Malthus that he got a clear view of the potency of natural selection.   This was in 1838—a year after he finished the first edition (it was not published until 1839), and seven years before the second edition was written (1845).   Thus the turning-point in the formation of his theory took place between the writing of the two editions.

I will first give a few passages which are practically the same in the two editions, and which are, therefore, chiefly of interest as illustrating his frame of mind in 1837.

The case of the two species of Molothrus (1st edit. p. 61 ; 2nd edit. p. 53) must have been one of the earliest instances noticed by him of the existence of representative species— a phenomenon which we know ('Autobiography,' p. 83) struck him deeply.   The discussion on introduced animals (1st edit. p. 139 ; 2nd edit. p. 120) shows how much he was impressed by the complicated interdependence of the inhabitants of a given area.

An analogous point of view is given in the discussion (1st edit. p. 98 ; 2nd edit. p. 85) of the mistaken belief that large animals require, for their support, a luxuriant vegetation ; the incorrectness of this view is illustrated by the comparison of the fauna of South Africa and South America, and the vegetation of the two continents. The interest of the discussion is that it shows clearly our *à priori* ignorance of the conditions of life suitable to any organism.

There is a passage which has been more than once quoted as bearing on the origin of his views. It is where he discusses the striking difference between the species of mice on the east and west of the Andes (1st edit. p. 399) : " Unless we suppose the same species to have been created in two different countries, we ought not to expect any closer similarity between the organic beings on the opposite sides of the Andes than on shores separated by a broad strait of the sea." In the 2nd edit. p. 327, the passage is almost verbally identical, and is practically the same.

There are other passages again which are more strongly evolutionary in the 2nd edit., but otherwise are similar to the corresponding passages in the 1st edition. Thus, in describing the blind Tuco-tuco (1st edit. p. 60 ; 2nd edit. p. 52), in the first edition he makes no allusion to what Lamarck might have thought, nor is the instance used as an example of modification, as in the edition of 1845.

A striking passage occurs in the 2nd edit. (p. 173) on the relationship between the " extinct edentata and the living sloths, ant-eaters, and armadillos."

" This wonderful relationship in the same continent between the dead and the living, will, I do not doubt, hereafter throw more light on the appearance of organic beings on our earth, and their disappearance from it, than any other class of facts."

This sentence does not occur in the 1st edit., but he was evidently profoundly struck by the disappearance of the gigantic forerunners of the present animals. The difference between the discussions in the two editions is most instructive. In both, our ignorance of the conditions of life is insisted on, but in the second edition, the discussion is made to lead up to a strong statement of the intensity of the struggle for life.

Then follows a comparison between rarity * and extinction,
which introduces the idea that the preservation and dominance
of existing species depend on the degree in which they are
adapted to surrounding conditions. In the first edition, he is
merely "tempted to believe in such simple relations as variation of climate and food, or introduction of enemies, or the
increased number of other species, as the cause of the succession of races."   But finally (1st edit.) he ends the chapter by
comparing the extinction of a species to the exhaustion and
disappearance of varieties of fruit-trees, as though he thought
that a mysterious term of life was impressed on each species
at its creation.

The difference of treatment of the Galapagos problem is of
some interest.   In the earlier book, the American type of the
productions of the islands is noticed, as is the fact that the
different islands possess forms specially their own, but the
importance of the whole problem is not so strongly put
forward.   Thus, in the first edition, he merely says :—

"This similarity of type between distant islands and continents, while the species are distinct, has scarcely been
sufficiently noticed.   The circumstance would be explained,
according to the views of some authors, by saying that the
creative power had acted according to the same law over a
wide area."—(1st edit. p. 474.)

This passage is not given in the second edition, and the
generalisations on geographical distribution are much wider
and fuller.   Thus he asks :—

"Why were their aboriginal inhabitants, associated . . . in
different proportions both in kind and number from those
on the Continent, and therefore acting on each other in a
different manner—why were they created on American types
of organisation?"—(2nd edit. p. 393.)

* In the second edition, p. 146, the destruction of Niata cattle by droughts is given as a good example of our ignorance of the causes of rarity or extinction. The passage does not occur in the first edition.

The same difference of treatment is shown elsewhere in this chapter. Thus the gradation in the form of beak presented by the thirteen allied species of finch is described in the first edition (p. 461) without comment. Whereas in the second edition (p. 380) he concludes :—

"One might really fancy that from an original paucity of birds in this Archipelago, one species has been taken and modified for different ends."

On the whole it seems to me remarkable that the difference between the two editions is not greater ; it is another proof of the author's caution and self-restraint in the treatment of his theory. After reading the second edition of the ' Journal,' we find with a strong sense of surprise how far developed were his views in 1837. We are enabled to form an opinion on this point from the note-books in which he wrote down detached thoughts and queries. I shall quote from the first note-book, completed between July 1837 and February 1838 : and this is the more worth doing, as it gives us an insight into the condition of his thoughts before the reading of Malthus. The notes are written in his most hurried style, so many words being omitted, that it is often difficult to arrive at the meaning. With a few exceptions (indicated by square brackets)* I have printed the extracts as written ; the punctuation, however, has been altered, and a few obvious slips corrected where it seemed necessary. The extracts are not printed in order, but are roughly classified.†

"Propagation explains why modern animals same type as extinct, which is law, almost proved."

"We can see why structure is common in certain countries

* In the extracts from the note-book ordinary brackets represent my father's parentheses.

† On the first page of the note-book, is written "Zoonomia" ; this seems to refer to the first few pages in which reproduction by gemma-tion is discussed, and where the "Zoonomia" is mentioned. Many pages have been cut out of the note-book, probably for use in writing the Sketch of 1844, and these would have no doubt contained the most interesting extracts.

when we can hardly believe necessary, but if it was necessary to one forefather, the result would be as it is. Hence antelopes at Cape of Good Hope ; marsupials at Australia."

" Countries longest separated greatest differences—if separated from immersage, possibly two distinct types, but each having its representatives—as in Australia."

" Will this apply to whole organic kingdom when our planet first cooled ? "

The two following extracts show that he applied the theory of evolution to the "whole organic kingdom" from plants to man.

" If we choose to let conjecture run wild, then animals, our fellow brethren in pain, disease, death, suffering and famine— our slaves in the most laborious works, our companions in our amusements—they may partake [of?] our origin in one common ancestor—we may be all melted together."

" The different intellects of man and animals not so great as between living things without thought (plants), and living things with thought (animals)."

The following extracts are again concerned with an *à priori* view of the probability of the origin of species by descent— "propagation," as he called it.

" The tree of life should perhaps be called the coral of life, base of branches dead ; so that passages cannot be seen."

" There never may have been grade between pig and tapir, yet from some common progenitor. Now if the intermediate ranks had produced infinite species, probably the series would have been more perfect."

At another place, speaking of intermediate forms, he says :—

" Cuvier objects to propagation of species by saying, why have not some intermediate forms been discovered between Palæotherium, Megalonyx, Mastodon, and the species now living ? Now according to my view (in S. America) parent of all Armadilloes might be brother to Megatherium—uncle now dead."

Speaking elsewhere of intermediate forms, he remarks :—
" Opponents will say—*show them me.* I will answer yes, if
you will show me every step between bulldog and greyhound."

Here we see that the case of domestic animals was already
present in his mind as bearing on the production of natural
species. The disappearance of intermediate forms naturally
leads up to the subject of extinction, with which the next
extract begins.

" It is a wonderful fact, horse, elephant, and mastodon,
dying out about same time in such different quarters.

" Will Mr. Lyell say that some [same ?] circumstance killed
it over a tract from Spain to South America ?—(Never.)

" They die, without they change, like golden pippins ; it is
a *generation of species* like generation *of individuals.*

" Why does individual die ? To perpetuate certain peculiarities (therefore adaptation), and obliterate accidental varieties,
and to accommodate itself to change (for, of course, change,
even in varieties, is accommodation). Now this argument
applies to species.

" If individual cannot propagate he has no issue—so with
species.

" If *species* generate other *species*, their race is not utterly cut
off :—like golden pippins, if produced by seed, go on—otherwise all die.

" The fossil horse generated, in South Africa, zebra—and
continued—perished in America.

" All animals of same species are bound together just like
buds of plants, which die at one time, though produced either
sooner or later. Prove animals like plants—trace gradation
between associated and non-associated animals—and the story
will be complete."

Here we have the view already alluded to of a term of life
impressed on a species.

But in the following note we get extinction connected with

unfavourable variation, and thus a hint is given of natural selection :—

"With respect to extinction, we can easily see that [a] variety of [the] ostrich (Petise), may not be well adapted, and thus perish out ; or, on the other hand, like Orpheus [a Galapagos bird], being favourable, many might be produced. This requires [the] principle that the permanent variations produced by confined breeding and changing circumstances are continued and produce[d] according to the adaptation of such circumstances, and therefore that death of species is a consequence (contrary to what would appear from America) of non-adaptation of circumstances."

The first part of the next extract has a similar bearing. The end of the passage is of much interest, as showing that he had at this early date visions of the far-reaching character of his speculations :—

"With belief of transmutation and geographical grouping, we are led to endeavour to discover *causes* of change ; the manner of adaptation (wish of parents??), instinct and structure becomes full of speculation and lines of observation. View of generation being condensation,* test of highest organisation intelligible . . . . My theory would give zest to recent and fossil comparative anatomy ; it would lead to the study of instincts, heredity, and mind-heredity, whole [of] metaphysics.

"It would lead to closest examination of hybridity, regeneration, causes of change in order to know what we have come from and to what we tend—to what circumstances favour crossing and what prevents it—this, and direct examination of direct passages of structure in species, might lead to laws of change, which would then be the main object of study, to guide our speculations."

The following two extracts have a similar interest ; the

---

* I imagine him to mean that each generation is " condensed " to a small number of the best organized individuals.

second is especially interesting, as it contains the germ of the concluding sentence of the 'Origin of Species' : *—

"Before the attraction of gravity discovered it might have been said it was as great a difficulty to account for the movement of all [planets] by one law, as to account for each separate one ; so to say that all mammalia were born from one stock, and since distributed by such means as we can recognise, may be thought to explain nothing.

"Astronomers might formerly have said that God fore-ordered each planet to move in its particular destiny. In the same manner God orders each animal created with certain forms in certain countries ; but how much more simple and sublime [a] power—let attraction act according to certain law, such are inevitable consequences—let animals be created, then by the fixed laws of generation, such will be their successors.

"Let the powers of transportal be such, and so will be the forms of one country to another—let geological changes go at such a rate, so will be the number and distribution of the species!!"

The three next extracts are of miscellaneous interest :—

"When one sees nipple on man's breast, one does not say some use, but sex not having been determined—so with useless wings under elytra of beetles—born from beetles with wings, and modified—if simple creation merely, would have been born without them."

"In a decreasing population at any one moment fewer closely related (few species of genera) ; ultimately few genera (for otherwise the relationship would converge sooner), and lastly, perhaps, some one single one. Will not this account

* 'Origin of Species' (edit. i.), p. 490 :—" There is a grandeur in this view of life, with its several powers, having been originally breathed into a few forms or into one ; and that whilst this planet has gone cycling on according to the fixed law of gravity, from so simple a beginning endless forms most beautiful and most wonderful have been, and are being evolved."

for the odd genera with few species which stand between great groups, which we are bound to consider the increasing ones?"

The last extract which I shall quote gives the germ of his theory of the relation between alpine plants in various parts of the world, in the publication of which he was forestalled by E. Forbes (see Vol. I. p. 88). He says, in the 1837 note-book, that alpine plants, "formerly descended lower, therefore [they are] species of lower genera altered, or northern plants."

When we turn to the Sketch of his theory, written in 1844 (still therefore before the second edition of the 'Journal' was completed), we find an enormous advance made on the note-book of 1837. The Sketch is in fact a surprisingly complete presentation of the argument afterwards familiar to us in the 'Origin of Species.' There is some obscurity as to the date of the short Sketch which formed the basis of the 1844 Essay. We know from his own words (Vol. I. p. 184), that it was in June 1842 that he first wrote out a short sketch of his views.* This statement is given with so much circumstance that it is almost impossible to suppose that it contains an error of date. It agrees also with the following extract from his Diary.

" 1842.   May 18th.   Went to Maer.

" June 15th to Shrewsbury, and on 18th to Capel Curig. During my stay at Maer and Shrewsbury (five years after commencement) wrote pencil-sketch of species theory."

Again in the introduction to the 'Origin,' p. 1, he writes, "after an interval of five years' work," [from 1837, i.e. in 1842,] "I allowed myself to speculate on the subject, and drew up some short notes."

Nevertheless in the letter signed by Sir C. Lyell and Sir J. D. Hooker, which serves as an introduction to the joint paper of Messrs. C. Darwin and A. Wallace on the 'Tendency

* This version I cannot find, and it was probably destroyed, like so much of his MS., after it had been enlarged and re-copied in 1844.

of Species to form Varieties,' * the essay of 1844 (extracts
from which form part of the paper) is said to have been
" sketched in 1839, and copied in 1844." This statement is
obviously made on the authority of a note written in my
father's hand across the Table of Contents of the 1844 Essay.
It is to the following effect : "This was sketched in 1839, and
copied out in full, as here written and read by you in 1844."
I conclude that this note was added in 1858, when the MS.
was sent to Sir J. D. Hooker (see Letter of June 29, 1858,
Vol. II. p. 119). There is also some further evidence on this side
of the question. Writing to Mr. Wallace (Jan. 25, 1859) my
father says :—"Every one whom I have seen has thought
your paper very well written and interesting. It puts my
extracts (written in 1839, now just twenty years ago !), which
I must say in apology were never for an instant intended for
publication, into the shade." The statement that the earliest
sketch was written in 1839 has been frequently made in
biographical notices of my father, no doubt on the authority
of the ' Linnean Journal,' but it must, I think, be considered
as erroneous. The error may possibly have arisen in this
way. In writing on the Table of Contents of the 1844 MS.
that it was sketched in 1839, I think my father may have
intended to imply that the framework of the theory was clearly
thought out by him at that date. In the Autobiography
(p. 88) he speaks of the time, " about 1839, when the theory
was clearly conceived," meaning, no doubt, the end of 1838
and beginning of 1839, when the reading of Malthus had
given him the key to the idea of natural selection. But this
explanation does not apply to the letter to Mr. Wallace ; and
with regard to the passage † in the 'Linnean Journal' it is
difficult to understand how it should have been allowed to

* 'Linn. Soc. Journal,' 1858,
p. 45.
† My father certainly saw the
proofs of the paper, for he added a
footnote apologising for the style of
the extracts, on the ground that the
" work was never intended for publication."

remain as it now stands, conveying, as it clearly does, the impression that 1839 was the date of his earliest written sketch. The sketch of 1844 is written in a clerk's hand, in two hundred and thirty-one pages folio, blank leaves being alternated with the MS. with a view to amplification. The text has been revised and corrected, criticisms being pencilled by himself on the margin. It is divided into two parts : I. "On the variation of Organic Beings under Domestication and in their Natural State." II. "On the Evidence favourable and opposed to the view that Species are naturally formed races descended from common Stocks." The first part contains the main argument of the 'Origin of Species.' It is founded, as is the argument of that work, on the study of domestic animals, and both the Sketch and the 'Origin' open with a chapter on variation under domestication and on artificial selection. This is followed, in both essays, by discussions on variation under nature, on natural selection, and on the struggle for life. Here, any close resemblance between the two essays with regard to arrangement ceases. Chapter III. of the Sketch, which concludes the first part, treats of the variations which occur in the instincts and habits of animals, and thus corresponds to some extent with Chapter VII. of the 'Origin' (1st edit.). It thus forms a complement to the chapters which deal with variation in structure. It seems to have been placed thus early in the Essay to prevent the hasty rejection of the whole theory by a reader to whom the idea of natural selection acting on instincts might seem impossible. This is the more probable, as the Chapter on Instinct in the 'Origin' is specially mentioned (Introduction, p. 5) as one of the "most apparent and gravest difficulties on the theory." Moreover the chapter in the Sketch ends with a discussion, "whether any particular corporeal structures . . . . . are so wonderful as to justify the rejection *primâ facie* of our theory." Under this heading comes the discussion of the eye, which in the 'Origin' finds its place in Chapter VI.

under " Difficulties on Theory." The second part seems to
have been planned in accordance with his favourite point of
view with regard to his theory. This is briefly given in a
letter to Dr. Asa Gray, November 11th, 1859: "I cannot
possibly believe that a false theory would explain so many
classes of facts, as I think it certainly does explain. On these
grounds I drop my anchor, and believe that the difficulties
will slowly disappear." On this principle, having stated the
theory in the first part, he. proceeds to show to what extent
various wide series of facts can be explained by its means.

Thus the second part of the Sketch corresponds roughly
to the nine concluding Chapters of the First Edition of the
' Origin.' But we must exclude Chapter VII. ('Origin') on
Instinct, which forms a chapter in the first part of the Sketch,
and Chapter VIII. ('Origin') on Hybridism, a subject treated
in the Sketch with 'Variation under Nature' in the first part.

The following list of the chapters of the second part of the
Sketch will illustrate their correspondence with the final
chapters of the ' Origin.'

Chapter I. "On the kind of intermediateness necessary,
and the number of such intermediate forms."

This includes a geological discussion, and corresponds to
parts of Chapters VI. and IX. of the ' Origin.'

Chapter II. " The gradual appearance and disappearance
of organic beings." Corresponds to Chapter X. of the
Origin.'

Chapter III. "Geographical Distribution." Corresponds to
Chapters XI. and XII. of the ' Origin.'

Chapter IV. "Affinities and Classification of Organic
beings."

Chapter V. " Unity of Type," Morphology, Embryology.

Chapter VI. Rudimentary Organs.

These three chapters correspond to Chapter XII. of the
' Origin.'

Chapter VII. Recapitulation and Conclusion. The final

sentence of the Sketch, which we saw in its first rough form in the Note Book of 1837, closely resembles the final sentence of the 'Origin,' much of it being identical. The 'Origin' is not divided into two "Parts," but we see traces of such a division having been present in the writer's mind, in this resemblance between the second part of the Sketch and the final chapters of the 'Origin.' That he should speak * of the chapters on transition, on instinct, on hybridism, and on the geological record, as forming a group, may be due to the division of his early MS. into two parts.

Mr. Huxley, who was good enough to read the Sketch at my request, while remarking that the "main lines of argument" and the illustrations employed are the same, points out that in the 1844 Essay, "much more weight is attached to the influence of external conditions in producing variation, and to the inheritance of acquired habits than in the 'Origin.'"

It is extremely interesting to find in the Sketch the first mention of principles familiar to us in the 'Origin of Species.' Foremost among these may be mentioned the principle of Sexual Selection, which is clearly enunciated. The important form of selèction known as "unconscious," is also given. Here also occurs a statement of the law that peculiarities tend to appear in the offspring at an age corresponding to that at which they occurred in the parent.

Professor Newton, who was so kind as to look through the 1844 Sketch, tells me that my father's remarks on the migration of birds, incidentally given in more than one passage, show that he had anticipated the views of some later writers.

With regard to the general style of the Sketch, it is not to be expected that it should have all the characteristics of the 'Origin,' and we do not, in fact, find that balance and control, that concentration and grasp, which are so striking in the work of 1859.

* 'Origin,' Introduction, p. 5.

In the Autobiography (Vol. I. p. 84) my father has stated what seemed to him the chief flaw of the 1844 Sketch; he had overlooked "one problem of great importance," the problem of the divergence of character. This point is discussed in the 'Origin of Species,' but, as it may not be familiar to all readers, I will give a short account of the difficulty and its solution. The author begins by stating that varieties differ from each other less than species, and then goes on: "Nevertheless, according to my view, varieties are species in process of formation. . . . . How then does the lesser difference between varieties become augmented into the greater difference between species." * He shows how an analogous divergence takes place under domestication where an originally uniform stock of horses has been split up into race-horses, dray-horses, &c., and then goes on to explain how the same principle applies to natural species. "From the simple circumstance that the more diversified the descendants from any one species become in structure, constitution, and habits, by so much will they be better enabled to seize on many and widely diversified places in the polity of nature, and so be enabled to increase in numbers."

The principle is exemplified by the fact that if on one plot of ground a single variety of wheat be sown, and on to another a mixture of varieties, in the latter case the produce is greater. More individuals have been able to exist because they were not all of the same variety. An organism becomes more perfect and more fitted to survive when by division of labour the different functions of life are performed by different organs. In the same way a species becomes more efficient and more able to survive when different sections of the species become differentiated so as to fill different stations.

In reading the Sketch of 1844, I have found it difficult to recognise, as a flaw in the Essay, the absence of any definite statement of the principle of divergence. Descent with

* 'Origin,' 1st edit. p. 111.

modification implies divergence, and we become so habituated to a belief in descent, and therefore in divergence, that we do not notice the absence of proof that divergence is in itself an advantage.   As shown in the Autobiography, my father in 1876 found it hardly credible that he should have overlooked the problem and its solution.

The following letter will be more in place here than its chronological position, since it shows what was my father's feeling as to the value of the Sketch at the time of its completion.]

C. Darwin to Mrs. Darwin.

Down, July 5, 1844.

. . . I have just finished my sketch of my species theory. If, as I believe, my theory in time be accepted even by one competent judge, it will be a considerable step in science.

I therefore write this in case of my sudden death, as my most solemn and last request, which I am sure you will consider the same as if legally entered in my will, that you will devote £400 to its publication, and further, will yourself, or through Hensleigh,* take trouble in promoting it.   I wish that my sketch be given to some competent person, with this sum to induce him to take trouble in its improvement and enlargement.   I give to him all my books on Natural History, which are either scored or have references at the end to the pages, begging him carefully to look over and consider such passages as actually bearing, or by possibility bearing, on this subject.   I wish you to make a list of all such books as some temptation to an editor.   I also request that you will hand over [to] him all those scraps roughly divided in eight or ten brown paper portfolios.   The scraps, with copied quotations from various works, are those which may aid my editor.   I also request that you, or some amanuensis, will aid

* Mr. H. Wedgwood.

in deciphering any of the scraps which the editor may think possibly of use. I leave to the editor's judgment whether to interpolate these facts in the text, or as notes, or under appendices. As the looking over the references and scraps will be a long labour, and as the *correcting* and enlarging and altering my sketch will also take considerable time, I leave this sum of £400 as some remuneration, and any profits from the work. I consider that for this the editor is bound to get the sketch published either at a publisher's or his own risk. Many of the scraps in the portfolios contain mere rude suggestions and early views, now useless, and many of the facts will probably turn out as having no bearing on my theory.

With respect to editors, Mr. Lyell would be the best if he would undertake it ; I believe he would find the work pleasant, and he would learn some facts new to him. As the editor must be a geologist as well as a naturalist, the next best editor would be Professor Forbes of London. The next best (and quite best in many respects) would be Professor Henslow. Dr. Hooker would be *very* good. The next, Mr. Strickland.* If none of these would undertake it, I would request you to consult with Mr. Lyell, or some other capable man for some editor, a geologist and naturalist. Should one other hundred pounds make the difference of procuring a good editor, I request earnestly that you will raise £500.

My remaining collections in Natural History may be given to any one or any museum where [they] would be accepted. . . .

[The following note seems to have formed part of the original letter, but may have been of later date :

"Lyell, especially with the aid of Hooker (and if any good zoological aid), would be best of all. Without an editor will pledge himself to give up time to it, it would be of no use paying such a sum.

---

* After Mr. Strickland's name comes the following sentence, which has been erased, but remains legible. "Professor Owen would be very good ; but I presume he would not undertake such a work."

"If there should be any difficulty in getting an editor who would go thoroughly into the subject, and think of the bearing of the passages marked in the books and copied out of scraps of paper, then let my sketch be published as it is, stating that it was done several years ago * and from memory without consulting any works, and with no intention of publication in its present form."

The idea that the Sketch of 1844 might remain, in the event of his death, as the only record of his work, seems to have been long in his mind, for in August 1854, when he had finished with the Cirripedes, and was thinking of beginning his "species work," he added on the back of the above letter, " Hooker by far best man to edit my species volume. August 1854."]

* The words " several years ago and," seem to have been added at a later date.

(19)

CHAPTER II.

THE GROWTH OF THE 'ORIGIN OF SPECIES.
LETTERS, 1843–1856.

[THE history of my father's life is told more completely in his correspondence with Sir J. D. Hooker than in any other series of letters; and this is especially true of the history of the growth of the 'Origin of Species.' This, therefore, seems an appropriate place for the following notes, which Sir Joseph Hooker has kindly given me. They give, moreover, an interesting picture of his early friendship with my father :—

"My first meeting with Mr. Darwin was in 1839, in Trafalgar Square. I was walking with an officer who had been his shipmate for a short time in the *Beagle* seven years before, but who had not, I believe, since met him. I was introduced; the interview was of course brief, and the memory of him that I carried away and still retain was that of a rather tall and rather broad-shouldered man, with a slight stoop, an agreeable and animated expression when talking, beetle brows, and a hollow but mellow voice; and that his greeting of his old acquaintance was sailor-like— that is, delightfully frank and cordial. I observed him well, for I was already aware of his attainments and labours, derived from having read various proof-sheets of his then unpublished 'Journal.' These had been submitted to Mr. (afterwards Sir Charles) Lyell by Mr. Darwin, and by him sent to his father, Ch. Lyell, Esq., of Kinnordy, who (being a very old friend of my father, and taking a kind interest in my projected career as a naturalist) had allowed me to peruse them. At this time

C 2

I was hurrying on my studies, so as to take my degree before volunteering to accompany Sir James Ross in the Antarctic Expedition, which had just been determined on by the Admiralty ; and so pressed for time was I, that I used to sleep with the sheets of the 'Journal' under my pillow, that I might read them between waking and rising. They impressed me profoundly, I might say despairingly, with the variety of acquirements, mental and physical, required in a naturalist who should follow in Darwin's footsteps, whilst they stimulated me to enthusiasm in the desire to travel and observe.

"It has been a permanent source of happiness to me that I knew so much of Mr. Darwin's scientific work so many years before that intimacy began which ripened into feelings as near to those of reverence for his life, works, and character as is reasonable and proper. It only remains to add to this little episode that I received a copy of the 'Journal' complete,—a gift from Mr. Lyell,—a few days before leaving England.

"Very soon after the return of the Antarctic Expedition my correspondence with Mr. Darwin began (December, 1843) by his sending me a long letter, warmly congratulating me on my return to my family and friends, and expressing a wish to hear more of the results of the expedition, of which he had derived some knowledge from private letters of my own (written to or communicated through Mr. Lyell). Then, plunging at once into scientific matters, he directed my attention to the importance of correlating the Fuegian Flora with that of the Cordillera and of Europe, and invited me to study the botanical collections which he had made in the Galapagos Islands, as well as his Patagonian and Fuegian plants.

"This led to me sending him an outline of the conclusions I had formed regarding the distribution of plants in the southern regions, and the necessity of assuming the destruction of considerable areas of land to account for the relations

of the flora of the so-called Antarctic Islands. I do not
suppose that any of these ideas were new to him, but they
led to an animated and lengthy correspondence full of
instruction."

Here follows the letter (1843) to Sir J. D. Hooker above
referred to.]

MY DEAR SIR,—I had hoped before this time to have had
the pleasure of seeing you and congratulating you on your
safe return from your long and glorious voyage. But as I
seldom go to London, we may not yet meet for some time—
without you are led to attend the Geological meetings.

I am anxious to know what you intend doing with all your
materials—I had so much pleasure in reading parts of some
of your letters, that I shall be very sorry if I, as one of the
public, have no opportunity of reading a good deal more.
I suppose you are very busy now and full of enjoyment:
how well I remember the happiness of my first few months
of England—it was worth all the discomforts of many a gale!
But I have run from the subject, which made me write, of
expressing my pleasure that Henslow (as he informed me
a few days since by letter) has sent to you my small collection of plants. You cannot think how much pleased I am,
as I feared they would have been all lost, and few as they are,
they cost me a good deal of trouble. There are a very few
notes, which I believe Henslow has got, describing the
habitats, &c., of some few of the more remarkable plants.
I paid particular attention to the Alpine flowers of Tierra del
Fuego, and I am sure I got every plant which was in flower
in Patagonia at the seasons when we were there. I have long
thought that some general sketch of the Flora of the point of
land, stretching so far into the southern seas, would be very
curious. Do make comparative remarks on the species allied
to the European species, for the advantage of botanical
ignoramuses like myself. It has often struck me as a curious

point to find out, whether there are many European genera in T. del Fuego which are not found along the ridge of the Cordillera; the separation in such case would be so enormous. Do point out in any sketch you draw up, what genera are American and what European, and how great the differences of the species are, when the genera are European, for the sake of the ignoramuses.

I hope Henslow will send you my Galapagos plants (about which Humboldt even expressed to me considerable curiosity) —I took much pains in collecting all I could. A Flora of this archipelago would, I suspect, offer a nearly parallel case to that of St. Helena, which has so long excited interest. Pray excuse this long rambling note, and believe me, my dear sir, yours very sincerely,

C. DARWIN.

Will you be so good as to present my respectful compliments to Sir W. Hooker.

[Referring to Sir J. D. Hooker's work on the Galapagos Flora, my father wrote in 1846:

" I cannot tell you how delighted and astonished I am at the results of your examination; how wonderfully they support my assertion on the differences in the animals of the different islands, about which I have always been fearful."

Again he wrote (1849) :—

" I received a few weeks ago your Galapagos papers,* and I have read them since being here. I really cannot express too strongly my admiration of the geographical discussion : to my judgment it is a perfect model of what such a paper should be ; it took me four days to read and think over. How interesting the Flora of the Sandwich Islands appears to be, how I wish there were materials for you to treat its

* These papers include the results of Sir J. D. Hooker's examination of my father's Galapagos plants, and were published by the Linnean Society in 1849.

flora as you have done the Galapagos. In the Systematic paper I was rather disappointed in not finding general remarks on affinities, structures, &c., such as you often give in conversation, and such as De Candolle and St. Hilaire introduced in almost all their papers, and which make them interesting even to a non-Botanist."

"Very soon afterwards [continues Sir J. D. Hooker] in a letter dated January 1844, the subject of the 'Origin of Species' was brought forward by him, and I believe that I was the first to whom he communicated his then new ideas on the subject, and which being of interest as a contribution to the history of Evolution, I here copy from his letter " :—]

C. Darwin to J. D. Hooker.

[January 11th, 1844.]

. . . Besides a general interest about the southern lands, I have been now ever since my return engaged in a very presumptuous work, and I know no one individual who would not say a very foolish one. I was so struck with the distribution of the Galapagos organisms, &c. &c., and with the character of the American fossil mammifers, &c. &c., that I determined to collect blindly every sort of fact, which could bear any way on what are species. I have read heaps of agricultural and horticultural books, and have never ceased collecting facts. At last gleams of light have come, and I am almost convinced (quite contrary to the opinion I started with) that species are not (it is like confessing a murder) immutable. Heaven forfend me from Lamarck nonsense of a "tendency to progression," "adaptations from the slow willing of animals," &c.! But the conclusions I am led to are not widely different from his; though the means of change are wholly so. I think I have found out (here's presumption!) the simple way by which species become exquisitely adapted to various ends. You will now groan, and think to yourself, "on what a man have I been wasting my time and writing to." I should, five years ago, have thought so. . . .

[The following letter written on February 23, 1844, shows that the acquaintanceship with Sir J. D. Hooker was then fast ripening into friendship. The letter is chiefly of interest as showing the sort of problems then occupying my father's mind :]

DEAR HOOKER,—I hope you will excuse the freedom of my address, but I feel that as co-circum-wanderers and as fellow labourers (though myself a very weak one) we may throw aside some of the old-world formality. . . . I have just finished a little volume on the volcanic islands which we visited. I do not know how far you care for dry simple geology, but I hope you will let me send you a copy. I suppose I can send it from London by common coach conveyance.

. . . I am going to ask you some *more* questions, though I dare say, without asking them, I shall see answers in your work, when published, which will be quite time enough for my purposes. First for the Galapagos, you will see in my Journal, that the Birds, though peculiar species, have a most obvious S. American aspect : I have just ascertained the same thing holds good with the sea-shells. Is it so with those plants which are peculiar to this archipelago ; you state that their numerical proportions are continental (is not this a very curious fact ?) but are they related in forms to S. America. Do you know of any other case of an archipelago, with the separate islands possessing distinct representative species ? I have always intended (but have not yet done so) to examine Webb and Berthelot on the Canary Islands for this object. Talking with Mr. Bentham, he told me that the separate islands of the Sandwich Archipelago possessed distinct representative species of the same genera of Labiatæ : would not this be worth your enquiry ? How is it with the

Azores ; to be sure the heavy western gales would tend to diffuse the same species over that group.

I hope you will (I dare say my hope is quite superfluous) attend to this general kind of affinity in isolated islands, though I suppose it is more difficult to perceive this sort of relation in plants, than in birds or quadrupeds, the groups of which are, I fancy, rather more confined. Can St. Helena be classed, though remotely, either with Africa or S. America ? From some facts, which I have collected, I have been led to conclude that the fauna of mountains are either remarkably similar (sometimes in the presence of the same species and at other times of same genera), or that they are remarkably dissimilar ; and it has occurred to me that possibly part of this peculiarity of the St. Helena and Galapagos floras may be attributed to a great part of these two Floras being mountain Floras. I fear my notes will hardly serve to distinguish much of the habitats of the Galapagos plants, but they may in some cases ; most, if not all, of the green, leafy plants come from the summits of the islands, and the thin brown leafless plants come from the lower arid parts : would you be so kind as to bear this remark in mind, when examining my collection.

I will trouble you with only one other question. In discussion with Mr. Gould, I found that in most of the genera of birds which range over the whole or greater part of the world, the individual species have wider ranges, thus the Owl is mundane, and many of the species have very wide ranges. So I believe it is with land and fresh-water shells—and I might adduce other cases. Is it not so with Cryptogamic plants ; have not most of the species wide ranges, in those genera which are mundane ? I do not suppose that the converse holds, viz.—that when a species has a wide range, its genus also ranges wide. Will you so far oblige me by occasionally thinking over this ? It would cost me vast trouble to get a list of mundane phanerogamic genera and

then search how far the species of these genera are apt to
range wide in their several countries; but you might occasionally, in the course of your pursuits, just bear this in mind,
though perhaps the point may long since have occurred to
you or other Botanists. Geology is bringing to light interesting facts, concerning the ranges of shells ; I think it is pretty
well established, that according as the geographical range of
a species is wide, so is its persistence and duration in time.
I hope you will try to grudge as little as you can the trouble
of my letters, and pray believe me very truly yours,

C. DARWIN.

P.S. I should feel extremely obliged for your kind offer of
the sketch of Humboldt ; I venerate him, and after having had
the pleasure of conversing with him in London, I shall still
more like to have any portrait of him.

[What follows is quoted from Sir J. D. Hooker's notes.

" The next act in the drama of our lives opens with personal
intercourse. This began with an invitation to breakfast with
him at his brother's (Erasmus Darwin's) house in Park Street ;
which was shortly afterwards followed by an invitation to
Down to meet a few brother Naturalists. In the short
intervals of good health that followed the long illnesses which
oftentimes rendered life a burthen to him, between 1844 and
1847, I had many such invitations, and delightful they were.
A more hospitable and more attractive home under every
point of view could not be imagined—of Society there were
most often Dr. Falconer, Edward Forbes, Professor Bell, and
Mr. Waterhouse—there were long walks, romps with the
children on hands and knees, music that haunts me still.
Darwin's own hearty manner, hollow laugh, and thorough
enjoyment of home life with friends ; strolls with him all
together, and interviews with us one by one in his study, to
discuss questions in any branch of biological or physical
knowledge that we had followed ; and which I at any rate

always left with the feeling that I had imparted nothing and
carried away more than I could stagger under.   Latterly, as
his health became more seriously affected, I was for days and
weeks the only visitor, bringing my work with me and
enjoying his society as opportunity offered.   It was an
established rule that he every day pumped me, as he called
it, for half an hour or so after breakfast in his study, when
he first brought out a heap of slips with questions botanical,
geographical, &c., for me to answer, and concluded by telling
me of the progress he had made in his own work, asking my
opinion on various points.   I saw no more of him till about
noon, when I heard his mellow ringing voice calling my
name under my window—this was to join him in his daily
forenoon walk round the sand-walk.*   On joining him I
found him in a rough grey shooting-coat in summer, and
thick cape over his shoulders in winter, and a stout staff in
his hand ; away we trudged through the garden, where there
was always some experiment to visit, and on to the sand-
walk, round which a fixed number of turns were taken, during
which our conversation usually ran on foreign lands and seas,
old friends, old books, and things far off to both mind and
eye.

" In the afternoon there was another such walk, after which
he again retired till dinner if well enough to join the family ;
if not, he generally managed to appear in the drawing-room,
where seated in his high chair, with his feet in enormous
carpet shoes, supported on a high stool—he enjoyed the
music or conversation of his family."

Here follows a series of letters illustrating the growth
of my father's views, and the nature of his work during this
period.]

* See Vol. I. p. 115.

C. Darwin to J. D. Hooker.

Down [1844].

... The conclusion, which I have come at is, that those areas, in which species are most numerous, have oftenest been divided and isolated from other areas, united and again divided ; a process implying antiquity and some changes in the external conditions. This will justly sound very hypothetical. I cannot give my reasons in detail ; but the most general conclusion, which the geographical distribution of all organic beings, appears to me to indicate, is that isolation is the chief concomitant or cause of the appearance of *new* forms (I well know there are some staring exceptions). Secondly, from seeing how often the plants and animals swarm in a country, when introduced into it, and from seeing what a vast number of plants will live, for instance in England, if kept *free from weeds, and native plants,* I have been led to consider that the spreading and number of the organic beings of any country depend less on its external features, than on the number of forms, which have been there originally created or produced. I much doubt whether you will find it possible to explain the number of forms by proportional differences of exposure ; and I cannot doubt if half the species in any country were destroyed or had not been created, yet that country would appear to us fully peopled. With respect to original creation or production of new forms, I have said that isolation appears the chief element. Hence, with respect to terrestrial productions, a tract of country, which had oftenest within the late geological periods subsided and been converted into islands, and reunited, I should expect to contain most forms.

But such speculations are amusing only to one's self, and in this case useless, as they do not show any direct line of observation : if I had seen how hypothetical [is] the little, which I

have unclearly written, I would not have troubled you with
the reading of it.   Believe me,—at last not hypothetically,

Yours very sincerely,

C. DARWIN.

C. Darwin to J. D. Hooker.

Down, 1844.

. . . I forget my last letter, but it must have been a very
silly one, as it seems I gave my notion of the number of
species being in great degree governed by the degree to
which the area had been often isolated and divided; I must
have been cracked to have written it, for I have no evidence,
without a person be willing to admit all my views, and then
it does follow; but in my most sanguine moments, all I
expect, is that I shall be able to show even to sound Naturalists, that there are two sides to the question of the immutability of species;—that facts can be viewed and grouped
under the notion of allied species having descended from
common stocks.   With respect to books on this subject, I
do not know of any systematical ones, except Lamarck's,
which is veritable rubbish; but there are plenty, as Lyell,
Pritchard, &c., on the view of the immutability.   Agassiz
lately has brought the strongest argument in favour of immutability.   Isidore G. St. Hilaire has written some good Essays,
tending towards the mutability-side, in the 'Suites à Buffon,'
entitled "Zoolog. Générale."   Is it not strange that the author
of such a book as the 'Animaux sans Vertèbres' should
have written that insects, which never see their eggs, should
will (and plants, their seeds) to be of particular forms, so as
to become attached to particular objects.   The other common
(specially Germanic) notion is hardly less absurd, viz. that
climate, food, &c., should make a Pediculus formed to climb
hair, or wood-pecker to climb trees.   I believe all these
absurd views arise from no one having, as far as I know,
approached the subject on the side of variation under domest-

ication, and having studied all that is known about domestication. I was very glad to hear your criticism on island-floras and on non-diffusion of plants : the subject is too long for a letter : I could defend myself to some considerable extent, but I doubt whether successfully in your eyes, or indeed in my own. . . .

C. Darwin to J. D. Hooker.

Down, [July, 1844.]

. . . I am now reading a wonderful book for facts on variation—Bronn, 'Geschichte der Natur.' It is stiff German : it forestalls me, sometimes I think delightfully, and sometimes cruelly. You will be ten times hereafter more horrified at me than at H. Watson. I hate arguments from results, but on my views of descent, really Natural History becomes a sublimely grand result-giving subject (now you may quiz me for so foolish an escape of mouth). . . . I must leave this letter till to-morrow, for I am tired ; but I so enjoy writing to you, that I must inflict a little more on you.

Have you any good evidence for absence of insects in small islands ? I found thirteen species in Keeling Atoll. Flies are good fertilizers, and I have seen a microscopic Thrips and a Cecidomya take flight from a flower in the direction of another with pollen adhering to them. In Arctic countries a bee seems to go as far N. as any flower. . . . . .

C. Darwin to J. D. Hooker.

Shrewsbury [September, 1845].

MY DEAR HOOKER,—I write a line to say that Cosmos * arrived quite safely (N.B. One sheet came loose in Pt. I.), and to thank you for your nice note. I have just begun the introduction, and groan over the style, which in such parts is full half the battle. How true many of the remarks are (*i.c.* as far as I can understand the wretched English) on the scenery ; it is an exact expression of one's own thoughts.

* A translation of Humboldt's ' Kosmos.'

I wish I ever had any books to lend you in return for the many you have lent me. . . .

All of what you kindly say about my species work does not alter one iota my long self-acknowledged presumption in accumulating facts and speculating on the subject of variation, without having worked out my due share of species. But now for nine years it has been anyhow the greatest amusement to me.

Farewell, my dear Hooker, I grieve more than you can well believe, over our prospect of so seldom meeting.

I have never perceived but one fault in you, and that you have grievously, viz. modesty; you form an exception to Sydney Smith's aphorism, that merit and modesty have no other connection, except in their first letter. Farewell,

C. DARWIN.

C. Darwin to L. Jenyns (Blomefield).

Down, Oct. 12th [1845].

MY DEAR JENYNS,—Thanks for your note. I am sorry to say I have not even the tail-end of a fact in English Zoology to communicate. I have found that even trifling observations require, in my case, some leisure and energy, both of which ingredients I have had none to spare, as writing my Geology thoroughly expends both. I had always thought that I would keep a journal and record everything, but in the way I now live I find I observe nothing to record. Looking after my garden and trees, and occasionally a very little walk in an idle frame of my mind, fills up every afternoon in the same manner. I am surprised that with all your parish affairs, you have had time to do all that which you have done. I shall be very glad to see your little work * (and

* Mr. Jenyns' 'Observations in Natural History.' It is prefaced by an Introduction on "Habits of observing as connected with the study of Natural History," and fol-lowed by a "Calendar of Periodic Phenomena in Natural History," with "Remarks on the importance of such Registers."

proud should I have been if I could have added a single fact
to it). My work on the species question has impressed me
very forcibly with the importance of all such works as your
intended one, containing what people are pleased generally
to call trifling facts. These are the facts which make one
understand the working or economy of nature. There is one
subject, on which I am very curious, and which perhaps you
may throw some light on, if you have ever thought on it ;
namely, what are the checks and what the periods of life,—
by which the increase of any given species is limited. Just
calculate the increase of any bird, if you assume that only
half the young are reared, and these breed : within the *natural*
(*i.e.* if free from accidents) life of the parents the number of
individuals will become enormous, and I have been much
surprised to think how great destruction *must* annually or
occasionally be falling on every species, yet the means and
period of such destruction is scarcely perceived by us.

I have continued steadily reading and collecting facts on
variation of domestic animals and plants, and on the question
of what are species. I have a grand body of facts, and I
think I can draw some sound conclusions. The general conclusions at which I have slowly been driven from a directly
opposite conviction, is that species are mutable, and that
allied species are co-descendants from common stocks. I
know how much I open myself to reproach for such a conclusion, but I have at least honestly and deliberately come to
it. I shall not publish on this subject for several years. At
present I am on the Geology of South America. I hope to
pick up from your book some facts on slight variations in
structure or instincts in the animals of your acquaintance.

Believe me, ever yours,

C. DARWIN.

C. Darwin to L. Jenyns.*

Down, [1845 ?].

MY DEAR JENYNS,—I am very much obliged to you for
the trouble you have taken in having written me so long
a note. The question of where, when, and how the check
to the increase of a given species falls appears to me par
ticularly interesting, and our difficulty in answering it shows
how really ignorant we are of the lives and habits of our most
familiar species. I was aware of the bare fact of old birds
driving away their young, but had never thought of the effect
you so clearly point out, of local gaps in number being thus
immediately filled up. But the original difficulty remains ; for
if your farmers had not killed your sparrows and rooks, what
would have become of those which now immigrate into your
parish ? in the middle of England one is too far distant from
the natural limits of the rook and sparrow to suppose that
the young are thus far expelled from Cambridgeshire. The
check must fall heavily at some time of each species' life ;
for, if one calculates that only half the progeny are reared
and bred, how enormous is the increase ! One has, however,
no business to feel so much surprise at one's ignorance, when
one knows how impossible it is without statistics to conjecture the duration of life and percentage of deaths to births
in mankind. If it could be shown that apparently the birds
of passage *which breed here* and increase, return in the succeeding years in about the same number, whereas those that
come here for their winter and non-breeding season annually,
come here with the same numbers, but return with greatly
decreased numbers, one would know (as indeed seems
probable) that the check fell chiefly on full-grown birds
in the winter season, and not on the eggs and very young
birds, which has appeared to me often the most probable
period. If at any time any remarks on this subject should

* Rev. L. Blomefield.

occur to you, I should be most grateful for the benefit of them.

With respect to my far distant work on species, I must have expressed myself with singular inaccuracy if I led you to suppose that I meant to say that my conclusions were inevitable. They have become so, after years of weighing puzzles, to myself *alone;* but in my wildest day-dream, I never expect more than to be able to show that there are two sides to the question of the immutability of species, *i.e.* whether species are *directly* created or by intermediate laws (as with the life and death of individuals). I did not approach the subject on the side of the difficulty in determining what are species and what are varieties, but (though why I should give you such a history of my doings it would be hard to say) from such facts as the relationship between the living and extinct mammifers in South America, and between those living on the Continent and on adjoining islands, such as the Galapagos. It occurred to me that a collection of all such analogous facts would throw light either for or against the view of related species being co-descendants from a common stock. A long searching amongst agricultural and horticultural books and people makes me believe (I well know how absurdly presumptuous this must appear) that I see the way in which new varieties become exquisitely adapted to the external conditions of life and to other surrounding beings. I am a bold man to lay myself open to being thought a complete fool, and a most deliberate one. From the nature of the grounds which make me believe that species are mutable in form, these grounds cannot be restricted to the closest-allied species ; but how far they extend I cannot tell, as my reasons fall away by degrees, when applied to species more and more remote from each other. Pray do not think that I am so blind as not to see that there are numerous immense difficulties in my notions, but they appear to me less than on the common view. I have

drawn up a sketch and had it copied (in 200 pages) of my
conclusions; and if I thought at some future time that you
would think it worth reading, I should, of course, be most
thankful to have the criticism of so competent a critic.
Excuse this very long and egotistical and ill-written letter,
which by your remarks you have led me into, and believe me,

Yours very truly,
C. DARWIN.

C. Darwin to L. Jenyns.

Down, Oct. 17th, 1846.

DEAR JENYNS,—I have taken a most ungrateful length
of time in thanking you for your very kind present of
your 'Observations.' But I happened to have had in hand
several other books, and have finished yours only a few days
ago. I found it very pleasant reading, and many of your
facts interested me much. I think I was more interested,
which is odd, with your notes on some of the lower animals
than on the higher ones. The introduction struck me as very
good ; but this is what I expected, for I well remember being
quite delighted with a preliminary essay to the first number
of the 'Annals of Natural History.' I missed one discussion,
and think myself ill-used, for I remember your saying you
would make some remarks on the weather and barometer,
as a guide for the ignorant in prediction. I had also hoped
to have perhaps met with some remarks on the amount of
variation in our common species. Andrew Smith once
declared he would get some hundreds of specimens of larks
and sparrows from all parts of Great Britain, and see whether,
with finest measurements, he could detect any proportional
variations in beaks or limbs, &c. This point interests me
from having lately been skimming over the absurdly opposite
conclusions of Gloger and Brehm; the one making half-a-
dozen species out of every common bird, and the other

D 2

turning so many reputed species into one. Have you ever
done anything of this kind, or have you ever studied Gloger's
or Brehm's works ? I was interested in your account of the
martins, for I had just before been utterly perplexed by
noticing just such a proceeding as you describe : I counted
seven, one day lately, visiting a single nest and sticking dirt
on the adjoining wall. I may mention that I once saw some
squirrels eagerly splitting those little semi-transparent
spherical galls on the back of oak-leaves for the maggot
within ; so that they are insectivorous. A *Cychrus rostratus*
once squirted into my eyes and gave me extreme pain ; and
I must tell you what happened to me on the banks of the
Cam, in my early entomological days : under a piece of
bark I found two *Carabi* (I forget which), and caught one in
each hand, when lo and behold I saw a sacred *Panagæus crux
major !* I could not bear to give up either of my *Carabi*, and
to lose *Panagæus* was out of the question ; so that in despair
I gently seized one of the *Carabi* between my teeth, when to
my unspeakable disgust and pain the little inconsiderate
beast squirted his acid down my throat, and I lost both *Carabi*
and *Panagæus* ! I was quite astonished to hear of a terrestrial *Planaria* ; for about a year or two ago I described in the
' Annals of Natural History' several beautifully coloured
terrestrial species of the Southern Hemisphere, and thought it
quite a new fact. By the way, you speak of a sheep with a
broken leg not having flukes : I have heard my father aver
that a fever, or any *serious accident*, as a broken limb, will
cause in a man all the intestinal worms to be evacuated.
Might not this possibly have been the case with the flukes in
their early state ?

I hope you were none the worse for Southampton ; * I wish
I had seen you looking rather fatter. I enjoyed my week
extremely, and it did me good. I missed you the last few
days, and we never managed to see much of each other ; but

* The meeting of the British Association.

there were so many people there, that I for one hardly saw anything of any one. Once again I thank you very cordially for your kind present, and the pleasure it has given me, and believe me,

Ever most truly yours,

C. DARWIN.

P.S.—I have quite forgotten to say how greatly interested I was with your discussion on the statistics of animals: when will Natural History be so perfect that such points as you discuss will be perfectly known about any one animal?

C. Darwin to J. D. Hooker.

Malvern, June 13 [1849].

. . . At last I am going to press with a small poor first-fruit of my confounded Cirripedia, viz. the fossil pedunculate cirripedia. You ask what effect studying species has had on my variation theories; I do not think much—I have felt some difficulties more. On the other hand, I have been struck (and probably unfairly from the class) with the variability of every part in some slight degree of every species. When the same organ is *rigorously* compared in many individuals, I always find some slight variability, and consequently that the diagnosis of species from minute differences is always dangerous. I had thought the same parts of the same species more resemble (than they do anyhow in Cirripedia) objects cast in the same mould. Systematic work would be easy were it not for this confounded variation, which, however, is pleasant to me as a speculatist, though odious to me as a systematist. Your remarks on the distinctness (so unpleasant to me) of the Himalayan Rubi, willows, &c., compared with those of northern [Europe?], &c., are very interesting; if my rude species-sketch had any *small* share in leading you to these observations, it has already done good and ample service, and may lay its bones in the earth in peace. I never heard anything so strange as Falconer's neglect of your letters ; I am extremely glad you are cordial with him again, though it must have cost you an effort. Falconer is a man one must love. . . . May you prosper in every way, my dear Hooker.

Your affectionate friend,

C. DARWIN.

C. Darwin to J. D. Hooker.

Down, Wednesday, [September, n. d.]

. . . Many thanks for your letter received yesterday, which, as always, set me thinking : I laughed at your attack at my stinginess in changes of level towards Forbes,* being so liberal towards myself; but I must maintain, that I have never let down or upheaved our mother-earth's surface, for the sake of explaining any one phenomenon, and I trust I have very seldom done so without some distinct evidence. So I must still think it a bold step (perhaps a very true one) to sink into the depths of ocean, within the period of existing species, so large a tract of surface. But there is no amount or extent of change of level, which I am not fully prepared to admit, but I must say I should like better evidence, than the identity of a few plants, which *possibly* (I do not say probably) might have been otherwise transported. Particular

---

* Edward Forbes, born in the Isle of Man 1815, died 1854. His best known work was his Report on the distribution of marine animals at different depths in the Mediterranean. An important memoir of his is referred to in my father's 'Autobiography,' p. 88. He held successively the posts of Curator to the Geological Society's Museum, and Professor of Natural History in the Museum of Practical Geology ; shortly before he died he was appointed Professor of Natural History in the University of Edinburgh. He seems to have impressed his contemporaries as a man of strikingly versatile and vigorous mind. The above allusion to changes of level refers to Forbes's tendency to explain the facts of geographical distribution by means of an active geological imagination.

thanks for your attempt to get me a copy of ' L'Espèce,' * and almost equal thanks for your criticisms on him : I rather misdoubted him, and felt not much inclined to take as gospel his facts. I find this one of my greatest difficulties with foreign authors, viz. judging of their credibility. How painfully (to me) true is your remark, that no one has hardly a right to examine the question of species who has not minutely described many. I was, however, pleased to hear from Owen (who is vehemently opposed to any mutability in species), that he thought it was a very fair subject, and that there was a mass of facts to be brought to bear on the question, not hitherto collected. My only comfort is (as I mean to attempt the subject), that I have dabbled in several branches of Natural History, and seen good specific men work out my species, and know something of geology (an indispensable union) ; and though I shall get more kicks than half-pennies, I will, life serving, attempt my work. Lamarck is the only exception, that I can think of, of an accurate describer of species, at least in the Invertebrate Kingdom, who has disbelieved in permanent species, but he in his absurd though clever work has done the subject harm, as has Mr. Vestiges, and, as (some future loose naturalist attempting the same speculations will perhaps say) has Mr. D. . . .

C. DARWIN.

C. Darwin to J. D. Hooker.

Down, September 25th [1853].

MY DEAR HOOKER,—I have read your paper with great interest ; it seems all very clear, and will form an admirable introduction to the New Zealand Flora, or to any Flora in the world. How few generalizers there are among systematists ;

* Probably Godron's essay, published by the Academy of Nancy in 1848–49, and afterwards as a separate book in 1859.

I really suspect there is something absolutely opposed to each
other and hostile in the two frames of mind required for
systematising and reasoning on large collections of facts.
Many of your arguments appear to me very well put, and,
as far as my experience goes, the candid way in which you
discuss the subject is unique.   The whole will be very useful
to me whenever I undertake my volume, though parts take
the wind very completely out of my sails ; it will be all nuts
to me . . . for I have for some time determined to give the
arguments on *both* sides (as far as I could), instead of arguing
on the mutability side alone.

   In my own Cirripedial work (by the way, thank you for
the dose of soft solder ; it does one—or at least me—a great
deal of good)—in my own work I have not felt conscious
that disbelieving in the mere *permanence* of species has made
much difference one way or the other ; in some few cases
(if publishing avowedly on the doctrine of non-permanence),
I should *not* have affixed names, and in some few cases
should have affixed names to remarkable varieties.   Certainly
I have felt it humiliating, discussing and doubting, and
examining over and over again, when in my own mind the
only doubt has been whether the form varied *to-day or
yesterday* (not to put too fine a point on it, as Snagsby * would
say).   After describing a set of forms as distinct species, tearing
up my MS., and making them one species, tearing that up
and making them separate, and then making them one
again (which has happened to me), I have gnashed my
teeth, cursed species, and asked what sin I had committed
to be so punished.   But I must confess that perhaps nearly
the same thing would have happened to me on any scheme
of work.

   I am heartily glad to hear your Journal† is so much
advanced ; how magnificently it seems to be illustrated !

---

* In 'Bleak House.'   † Sir J. D. Hooker's ' Himalayan Journal.'

An ' *Oriental Naturalist*,' with lots of imagination and not too much regard to facts, is just the man to discuss species! I think your title of ' A Journal of a Naturalist in the East ' very good ; but whether " in the Himalaya" would not be better, I have doubted, for the East sounds rather vague. . . .

C. Darwin to J. D. Hooker.

[1853.]

MY DEAR HOOKER,—I have no remarks at all worth sending you, nor, indeed, was it likely that I should, considering how perfect and elaborated an essay it is.* As far as my judgment goes, it is the most important discussion on the points in question ever published. I can say no more. I agree with almost everything you say ; but I require much time to digest an essay of such quality. It almost made me gloomy, partly from feeling I could not answer some points which theoretically I should have liked to have been different, and partly from seeing *so far better done* than I could have done, discussions on some points which I had intended to have taken up. . . .

I much enjoyed the slaps you have given to the provincial species-mongers. I wish I could have been of the slightest use : I have been deeply interested by the whole essay, and congratulate you on having produced a memoir which I believe will be memorable. I was deep in it when your most considerate note arrived, begging me not to hurry. I thank Mrs. Hooker and yourself most sincerely for your wish to see me. I will not let another summer pass without seeing you at Kew, for indeed I should enjoy it much. . . .

You do me really more honour than I have any claim to, putting me in after Lyell on ups and downs. In a year or two's time, when I shall be at my species book (if I do

* ' New Zealand Flora,' 1853.

not break down), I shall gnash my teeth and abuse you for having put so many hostile facts so confoundedly well.

Ever yours affectionately,

C. DARWIN.

C. Darwin to J. D. Hooker.

Down, March 26th [1854].

MY DEAR HOOKER,—I had hoped that you would have had a little breathing-time after your Journal, but this seems to be very far from the case ; and I am the more obliged (and somewhat contrite) for the long letter received this morning, *most* juicy with news and *most* interesting to me in many ways. I am very glad indeed to hear of the reforms, &c., in the Royal Society. With respect to the Club,* I am deeply interested ; only two or three days ago, I was regretting to my wife, how I was letting drop and being dropped by nearly all my acquaintances, and that I would endeavour to go oftener to London ; I was not then thinking of the Club, which, as far as any one thing goes, would answer my exact object in keeping up old and making some new acquaintances. I will therefore come up to London for every (with rare exceptions) Club-day, and then my head, I think, will allow me on an average to go to every other meeting. But it is

* The Philosophical Club, to which my father was elected (as Professor Bonney is good enough to inform me) on April 24, 1854. He resigned his membership in 1864. The Club was founded in 1847. The number of members being limited to 47, it was proposed to christen it "the Club of 47," but the name was never adopted. The nature of the Club may be gathered from its first rule : "The purpose of the Club is to promote as much as possible the scientific objects of the Royal Society ; to facilitate intercourse between those Fellows who are actively engaged in cultivating the various branches of Natural Science, and who have contributed to its progress ; to increase the attendance at the evening meetings, and to encourage the contribution and discussion of papers." The Club met for dinner at 6, and the chair was to be quitted at 8.15, it being expected that members would go to the Royal Society. Of late years the dinner has been at 6.30, the Society meeting in the afternoon.

grievous how often any change knocks me up. I will further pledge myself, as I told Lyell, to resign after a year, if I did not attend pretty often, so that I should *at worst* encumber the Club temporarily. If you can get me elected, I certainly shall be very much pleased. Very many thanks for answers about Glaciers. I am very glad to hear of the second Edit.* so very soon ; but am not surprised, for I have heard of several, in our small circle, reading it with very much pleasure. I shall be curious to hear what Humboldt will say : it will, I should think, delight him, and meet with more praise from him than any other book of Travels, for I cannot remember one, which has so many subjects in common with him. What a wonderful old fellow he is. . . . . By the way, I hope, when you go to Hitcham,† towards the end of May, you will be forced to have some rest. I am grieved to hear that all the bad symptoms have not left Henslow ; it is so strange and new to feel any uneasiness about his health. I am particularly obliged to you for sending me Asa Gray's letter ; how very pleasantly he writes. To see his and your caution on the species-question ought to overwhelm me in confusion and shame ; it does make me feel deuced uncomfortable. . . . It is delightful to hear all that he says on Agassiz : how very singular it is that so *eminently* clever a man, with such *immense* knowledge on many branches of Natural History, should write as he does. Lyell told me that he was so delighted with one of his (Agassiz') lectures on progressive development, &c. &c., that he went to him afterwards and told him, "that it was so delightful, that he could not help all the time wishing it was true." I seldom see a Zoological paper from North America, without observing the impress of Agassiz' doctrines,—another proof, by the way, of how great a man he is. I was pleased and surprised to see A. Gray's remarks on crossing, obliterating varieties, on which, as you know, I have been collecting facts for these dozen years.

* Of the Himalayan Journal.   † Henslow's living.

How awfully flat I shall feel, if, when I get my notes together
on species, &c. &c., the whole thing explodes like an empty
puff-ball. Do not work yourself to death.

Ever yours most truly,

C. DARWIN.

C. Darwin to J. D. Hooker.

Down, Nov. 5th [1854].

MY DEAR HOOKER,—I was delighted to get your note
yesterday. I congratulate you very heartily,* and whether
you care much or little, I rejoice to see the highest scientific
judgment-court in Great Britain recognise your claims. I do
hope Mrs. Hooker is pleased, and E. desires me particularly
to send her cordial congratulations. . . . I pity you from the
very bottom of my heart about your after-dinner speech,
which I fear I shall not hear. Without you have a very
much greater soul than I have (and I believe that you have),
you will find the medal a pleasant little stimulus ; when work
goes badly, and one ruminates that all is vanity, it is pleasant
to have some tangible proof, that others have thought something of one's labours.

Good-bye, my dear Hooker, I can assure [you] that we
both most truly enjoyed your and Mrs. Hooker's visit here.
Farewell.

My dear Hooker, your sincere friend,

C. DARWIN.

C. Darwin to J. D. Hooker.

March 7 [1855].

. . . I have just finished working well at Wollaston's †
' Insecta Maderensia ' : it is an *admirable* work. There is a

* On the award to him of the
Royal Society's Medal.
† Thomas Vernon Wollaston,
born March 9, 1821 ; died Jan. 4,

1878. His health forcing him
in early manhood to winter in
the south, he devoted himself to
a study of the Coleoptera of

very curious point in the astounding proportion of Coleoptera
that are apterous; and I think I have guessed the reason,
viz. that powers of flight would be injurious to insects inhabiting a confined locality, and expose them to be blown to the
sea : to test this, I find that the insects inhabiting the Dezerte
Grande, a quite small islet, would be still more exposed to
this danger, and here the proportion of apterous insects is
even considerably greater than on Madeira proper. Wollaston
speaks of Madeira and the other Archipelagoes as being
"sure and certain witnesses of Forbes' old continent," and of
course the Entomological world implicitly follows this view.
But to my eyes it would be difficult to imagine facts more
opposed to such a view. It is really disgusting and humiliating to see directly opposite conclusions drawn from the
same facts.

I have had some correspondence with Wollaston on this
and other subjects, and I find that he coolly assumes, (1) that
formerly insects possessed greater migratory powers than
now, (2) that the old land was *specially* rich in centres of
creation, (3) that the uniting land was destroyed before the
special creations had time to diffuse, and (4) that the land
was broken down before certain families and genera had
time to reach from Europe or Africa the points of land in
question. Are not these a jolly lot of assumptions? and yet
I shall see for the next dozen or score of years Wollaston

Madeira, the Cape de Verdes,
and St. Helena, whence he deduced
evidence in support of the belief
in the submerged continent of
'Atlantis.' In an obituary notice
by Mr. Rye ('Nature,' 1878) he
is described as working persis-
tently "upon a broad conception of
the science to which he was devoted," while being at the same
time "accurate, elaborate, and
precise *ad punctum*, and naturally

of a minutely critical habit." His
first scientific paper was written
when he was an undergraduate at
Jesus College, Cambridge. While
at the University, he was an Asso-
ciate and afterwards a Member of
the Ray Club : this is a small
society which still meets once a
week, and where the undergraduate
members, or Associates, receive
much kindly encouragement from
their elders.

quoted as proving the former existence of poor Forbes' Atlantis.

I hope I have not wearied you, but I thought you would like to hear about this book, which strikes me as *excellent* in its facts, and the author a most nice and modest man.

Most truly yours,
C. DARWIN.

C. Darwin to W. D. Fox.

Down, March 19th [1855].

MY DEAR FOX,—How long it is since we have had any communication, and I really want to hear how the world goes with you; but my immediate object is to ask you to observe a point for me, and as I know now you are a very busy man with too much to do, I shall have a good chance of your doing what I want, as it would be hopeless to ask a quite idle man. As you have a Noah's Ark, I do not doubt that you have pigeons. (How I wish by any chance they were fantails!) Now what I want to know is, at what age nestling pigeons have their tail feathers sufficiently developed to be counted. I do not think I ever saw a young pigeon. I am hard at work at my notes collecting and comparing them, in order in some two or three years to write a book with all the facts and arguments, which I can collect, *for and versus* the immutability of species. I want to get the young of our domestic breeds, to see how young, and to what degree the differences appear. I must either breed myself (which is no amusement but a horrid bore to me) the pigeons or buy their young; and before I go to a seller, whom I have heard of from Yarrell, I am really anxious to know something about their development, not to expose my excessive ignorance, and therefore be excessively liable to be cheated and gulled. With respect to the *one* point of the tail feathers, it is of course in relation to the wonderful development of tail feathers in the adult fantail. If you had any breed of poultry pure, I

would beg a chicken with exact age stated, about a week or fortnight old! to be sent in a box by post, if you could have the heart
to kill one ; and secondly, would let me pay postage . . . Indeed,
I should be very glad to have a nestling common pigeon sent,
for I mean to make skeletons, and have already just begun
comparing wild and tame ducks. And I think the results
rather curious,* for on weighing the several bones very carefully, when perfectly cleaned the proportional weights of the
two have greatly varied, the foot of the tame having largely
increased. How I wish I could get a little wild duck of a
week old, but that I know is almost impossible.

With respect to ourselves, I have not much to say ; we
have now a terribly noisy house with the whooping cough,
but otherwise are all well. Far the greatest fact about myself
is that I have at last quite done with the everlasting barnacles.
At the end of the year we had two of our little boys very ill
with fever and bronchitis, and all sorts of ailments. Partly
for amusement, and partly for change of air, we went to
London and took a house for a month, but it turned out
a great failure, for that dreadful frost just set in when we
went, and all our children got unwell, and E. and I had
coughs and colds and rheumatism nearly all the time. We
had put down first on our list of things to do, to go and
see Mrs. Fox, but literally after waiting some time to see
whether the weather would not improve, we had not a day
when we both could go out.

I do hope before very long you will be able to manage
to pay us a visit. Time is slipping away, and we are
getting oldish. Do tell us about yourself and all your large
family.

I know you will help me *if you can* with information

* " I have just been testing practically what disuse does in reducing
parts ; I have made skeleton of
wild and tame duck (oh, the smell
of well-boiled, high duck ! !) and I
find the tame-duck wing ought, according to scale of wild prototype,
to have its two wings 360 grains in
weight, but it has it only 317."—
A letter to Sir J. D. Hooker, 1855.

about the young pigeons; and anyhow do write before very
long.

My dear Fox, your sincere old friend,
C. DARWIN.

P.S.—Amongst all sorts of odds and ends, with which I am
amusing myself, I am comparing the seeds of the variations
of plants. I had formerly some wild cabbage seeds, which I
gave to some one, was it to you? It is a *thousand* to one it
was thrown away, if not I should be very glad of a pinch of it.

[The following extract from a letter to Mr. Fox (March 27th,
1855) refers to the same subject as the last letter, and gives
some account of the "species work:" "The way I shall kill
young things will be to put them under a tumbler glass with a
teaspoon of ether or chloroform, the glass being pressed down
on some yielding surface, and leave them for an hour or two;
young have such power of revivification. (I have thus killed
moths and butterflies.) The best way would be to send them
as you procure them, in pasteboard chip-boxes by post, on
which you could write and just tie up with string; and you will
really make me happier by allowing me to keep an account
of postage, &c. Upon my word I can hardly believe that
any one could be so good-natured as to take such trouble
and do such a very disagreeable thing as kill babies; and I
am very sure I do not know one soul who, except yourself,
would do so. I am going to ask one thing more; should
old hens of any above poultry (not duck) die or become so
old as to be *useless*, I wish you would send her to me per
rail, addressed to 'C. Darwin, care of Mr. Acton, Post-office,
Bromley, Kent.' Will you keep this address? as shortest
way for parcels. But I do not care so much for this, as I
could buy the old birds dead at Baily's to make skeletons.
I should have written at once even if I had not heard from
you, to beg you not to take trouble about pigeons, for Yarrell
has persuaded me to attempt it, and I am now fitting up a

place, and have written to Baily about prices, &c. &c. *Some-time* (when you are better) I should like very much to hear a little about your " Little Call Duck "; why so called? And where you got it? and what it is like? . . . I was so ignorant I did not even know there were three varieties of Dorking fowl: how do they differ? . . .

I forget whether I ever told you what the object of my present work is,—it is to view all facts that I can master (eheu, eheu, how ignorant I find I am) in Natural History (as on geographical distribution, palæontology, classification, hybridism, domestic animals and plants, &c. &c. &c.) to see how far they favour or are opposed to the notion that wild species are mutable or immutable: I mean with my utmost power to give all arguments and facts on both sides. I have a *number* of people helping me in every way, and giving me most valuable assistance; but I often doubt whether the subject will not quite overpower me.

So much for the quasi-business part of my letter. I am very very sorry to hear so indifferent an account of your health: with your large family your life is very precious, and I am sure with all your activity and goodness it ought to be a happy one, or as happy as can reasonably be expected with all the cares of futurity on one.

One cannot expect the present to be like the old Crux-major days at the foot of those noble willow stumps, the memory of which I revere. I now find my little entomology, which I wholly owe to you, comes in very useful. I am very glad to hear that you have given yourself a rest from Sunday duties. How much illness you have had in your life! Farewell, my dear Fox. I assure you I thank you heartily for your proffered assistance."]

C. Darwin to W. D. Fox.

Down, May 7th [1855].

MY DEAR FOX,—My correspondence has cost you a deal of trouble, though this note will not. I found yours on my return

home on Saturday after a week's work in London. Whilst
there I saw Yarrell, who told me he had carefully examined
all points in the Call Duck, and did not feel any doubt
about it being specifically identical, and that it had crossed
freely with common varieties in St. James's Park. I should
therefore be very glad for a seven-days' duckling and for one
of the old birds, should one ever die a natural death. Yarrell
told me that Sabine had collected forty varieties of the
common duck ! ... Well, to return to business ; nobody, I am
sure, could fix better for me than you the characteristic age of
little chickens ; with respect to skeletons, I have feared it
would be impossible to make them, but I suppose I shall be
able to measure limbs, &c., by feeling the joints. What you
say about old cocks just confirms what I thought, and I will
make my skeletons of old cocks. Should an old wild turkey
ever die, please remember me ; I do not care for a baby turkey, nor for a mastiff. Very many thanks for your offer. I
have puppies of bull-dogs and greyhound in salt, and I have
had cart-horse and race-horse young colts carefully mea-
sured. Whether I shall do any good I doubt. I am getting
out of my depth. Most truly yours,

C. DARWIN.

[An extract from a letter to Mr. Fox may find a place
here, though of a later date, viz. July, 1855 :

"Many thanks for the seven days old white Dorking, and
for the other promised ones. I am getting quite 'a chamber
of horrors ;' I appreciate your kindness even more than
before, for I have done the black deed and murdered an
angelic little fantail, and a pouter at ten days old. I tried
chloroform and ether for the first, and though evidently a
perfectly easy death, it was prolonged ; and for the second I
tried putting lumps of cyanide of potassium in a very large
damp bottle, half an hour before putting in the pigeon,

and the prussic acid gas thus generated was very quickly
fatal."

A letter to Mr. Fox (May 23rd, 1855) gives the first
mention of my father's laborious piece of work on the
breeding of pigeons :

" I write now to say that I have been looking at some of
our mongrel chickens, and I should say *one week old* would
do very well. The chief points which I am, and have been
for years, very curious about, is to ascertain whether the
young of our domestic breeds differ as much from each other
as do their parents, and I have no faith in anything short
of actual measurement and the Rule of Three. I hope and
believe I am not giving so much trouble without a motive of
sufficient worth.   I have got my fantails and pouters (choice
birds, I hope, as I paid 20*s.* for each pair from Baily) in a
grand cage and pigeon-house, and they are a decided amusement to me, and delight to H."

In the course of my father's pigeon-fancying enterprise he
necessarily became acquainted with breeders, and was fond of
relating his experiences as a member of the Columbarian
and Philoperistera Clubs, where he met the purest enthusiasts
of the " fancy," and learnt much of the mysteries of their art.
In writing to Mr. Huxley some years afterwards, he quotes
from a book on Pigeons by Mr. J. Eaton, in illustration of
the " extreme attention and close observation " necessary to
be a good fancier.

" In his [Mr. Eaton's] treatise, devoted to the Almond
Tumbler *alone*, which is a sub-variety of the short-faced
variety, which is a variety of the Tumbler, as that is of the
Rock-pigeon, Mr. Eaton says : ' There are some of the
young fanciers who are over-covetous, who go for all the
five properties at once (*i.e.* the five characteristic points
which are mainly attended to,—C. D.), they have their reward

by getting nothing.' In short, it is almost beyond the human intellect to attend to *all* the excellencies of the Almond Tumbler !

" To be a good breeder, and to succeed in improving any breed, beyond everything enthusiasm is required. Mr. Eaton has gained lots of prizes, listen to him.

" ' If it was possible for noblemen and gentlemen to know the amazing amount of solace and pleasure derived from the Almond Tumbler, when they begin to understand their (*i.e.* the tumbler's) properties, I should think that scarce any nobleman or gentleman would be without their aviaries of Almond Tumblers.' "

My father was fond of quoting this passage, and always with a tone of fellow-feeling for the author, though, no doubt, he had forgotten his own wonderings as a child that " every gentleman did not become an ornithologist." — ('Autobiography,' p. 35.)

To Mr. W. B. Tegetmeier, the well-known writer on poultry, &c., he was indebted for constant advice and co-operation. Their correspondence began in 1855, and lasted to 1881, when my father wrote : " I can assure you that I often look back with pleasure to the old days when I attended to pigeons, fowls, &c., and when you gave me such valuable assistance. I not rarely regret that I have had so little strength that I have not been able to keep up old acquaintances and friendships." My father's letters to Mr. Tegetmeier consist almost entirely of series of questions relating to the different breeds of fowls, pigeons, &c., and are not, therefore, interesting. In reading through the pile of letters, one is much struck by the diligence of the writer's search for facts, and it is made clear that Mr. Tegetmeier's knowledge and judgment were completely trusted and highly valued by him. Numerous phrases, such as " your note is a mine of wealth to me," occur, expressing his sense of the value of Mr. Tegetmeier's help, as well as words expressing his warm appreciation of Mr. Tegetmeier's unstinting zeal and kindness, or his "pure and disinterested love of science." On the subject of hive-bees and their combs, Mr. Tegetmeier's help was also valued by my father, who wrote, "your paper on 'Bees-cells,' read before the British Association, was highly useful and suggestive to me."

To work out the problems on the Geographical Distributions of animals and plants on evolutionary principles, he had to study the means by which seeds, eggs, &c., can be transported across wide spaces of ocean. It was this need which gave an interest to the class of experiment to which the following letters allude.]

C. Darwin to W. D. Fox.

Down, May 17th [1855].

MY DEAR FOX,—You will hate the very sight of my handwriting; but after this time I promise I will ask for nothing more, at least for a long time. As you live on sandy soil, have you lizards at all common? If you have, should you think it too ridiculous to offer a reward for me for lizard's eggs to the boys in your school; a shilling for every halfdozen, or more if rare, till you got two or three dozen and send them to me? If snake's eggs were brought in mistake it would be very well, for I want such also; and we have neither lizards nor snakes about here. My object is to see whether such eggs will float on sea water, and whether they will keep alive thus floating for a month or two in my cellar. I am trying experiments on transportation of all organic beings that I can; and lizards are found on every island, and therefore I am very anxious to see whether their eggs stand sea water. Of course this note need not be answered, without, by a strange and favourable chance, you can some day answer it with the eggs. Your most troublesome friend,

C. DARWIN.

C. Darwin to J. D. Hooker.

April 13th [1855].

. . . I have had one experiment some little time in progress which will, I think, be interesting, namely, seeds in salt water, immersed in water of 32°–33°, which I have and shall long have, as I filled a great tank with snow. When I wrote last I was going to triumph over you, for my experiment had in a slight degree succeeded ; but this, with infinite baseness, I did not tell, in hopes that you would say that you would eat all the plants which I could raise after immersion. It is very aggravating that I cannot in the least remember what you did formerly say that made me think you scoffed at the experiments vastly; for you now seem to view the experiment like a good Christian. I have in small bottles out of doors, exposed to variation of temperature, cress, radish, cabbages, lettuces, carrots, and celery, and onion seed—four great families. These, after immersion for exactly one week, have all germinated, which I did not in the least expect (and thought how you would sneer at me) ; for the water of nearly all, and of the cress especially, smelt very badly, and the cress seed emitted a wonderful quantity of mucus (the 'Vestiges' would have expected them to turn into tadpoles), so as to adhere in a mass ; but these seeds germinated and grew splendidly. The germination of all (especially cress and lettuces) has been accelerated, except the cabbages, which have come up very irregularly, and a good many, I think, dead. One would have thought, from their native habitat, that the cabbage would have stood well. The Umbelliferæ and onions seem to stand the salt well. I wash the seed before planting them. I have written to the *Gardeners' Chronicle,** though I doubt whether it was worth

* A few words asking for information. The results were published in the 'Gardeners' Chronicle,' May 26, Nov. 24, 1855. In the same year (p. 789) he sent a P.S. to his former paper, correcting a misprint and adding a few words on the seeds of the Leguminosæ. A fuller paper

while. If my success seems to make it worth while, I will send a seed list, to get you to mark some different classes of seeds. To-day I replant the same seeds as above after fourteen days' immersion. As many sea-currents go a mile an hour, even in a week they might be transported 168 miles ; the Gulf Stream is said to go fifty and sixty miles a day. So much and too much on this head; but my geese are always swans. . . .

C. Darwin to J. D. Hooker.

[April 14th, 1855.]

. . . You are a good man to confess that you expected the cress would be killed in a week, for this gives me a nice little triumph. The children at first were tremendously eager, and asked me often, "whether I should beat Dr. Hooker!" The cress and lettuce have just vegetated well after twenty-one days' immersion. But I will write no more, which is a great virtue in me ; for it is to me a very great pleasure telling you everything I do.

. . . If you knew some of the experiments (if they may be so called) which I am trying, you would have a good right to sneer, for they are so *absurd* even in *my* opinion that I dare not tell you.

Have not some men a nice notion of experimentising ? I have had a letter telling me that seeds *must* have *great* power of resisting salt water, for otherwise how could they get to islands ? This is the true way to solve a problem !

C. Darwin to J. D. Hooker.

Down, [1855.]

MY DEAR HOOKER,—You have been a very good man to exhale some of your satisfaction in writing two notes to me ;

---

on the germination of seeds after treatment in salt water, appeared in the ' Linnean Soc. Journal,' 1857, p. 130.

you could not have taken a better line, in my opinion ; but as
for showing your satisfaction in confounding my experiments,
I assure you I am quite enough confounded—those horrid
seeds, which, as you truly observe, if they sink they won't
float.

I have written to Scoresby and have had a rather dry
answer, but very much to the purpose, and giving me no
hopes of any law unknown to me which might arrest their
everlasting descent into the deepest depths of the ocean.   By
the way it was very odd, but I talked to Col. Sabine for half
an hour on the subject, and could not make him see with
respect to transportal the difficulty of the sinking question !
The bore is, if the confounded seeds will sink, I have been
taking all this trouble in salting the ungrateful rascals for
nothing.

Everything has been going wrong with me lately ; the fish
at the Zoolog. Soc. ate up lots of soaked seeds, and in
imagination they had in my mind been swallowed, fish and
all, by a heron, had been carried a hundred miles, been
voided on the banks of some other lake and germinated
splendidly, when lo and behold, the fish ejected vehemently,
and with disgust equal to my own, *all* the seeds from their
mouths.*

But I am not going to give up the floating yet: in first
place I must try fresh seeds, though of course it seems far
more probable that they will sink ; and secondly, as a last
resource, I must believe in the pod or even whole plant or
branch being washed into the sea ; with floods and slips and

* In describing these troubles to
Mr. Fox, my father wrote :—"All
nature is perverse and will not do
as I wish it ; and just at present I
wish I had my old barnacles to
work at, and nothing new." The
experiment ultimately succeeded,
and he wrote to Sir J. Hooker :—
"I find fish will greedily eat seeds
of aquatic grasses, and that millet-
seed put into fish and given to a
stork, and then voided, will germi-
nate. So this is the nursery rhyme
of 'this is the stick that beats the
pig,' &c. &c."

earthquakes ; this must continually be happening, and if kept wet, I fancy the pods, &c. &c., would not open and shed their seeds. Do try your Mimosa seed at Kew.

I had intended to have asked you whether the *Mimosa scandens* and *Guilandina bonduc* grows at Kew, to try fresh seeds. R. Brown tells me he believes four W. Indian seeds have been washed on shores of Europe. I was assured at Keeling Island that seeds were not rarely washed on shore : so float they must and shall! What a long yarn I have been spinning.

If you have several of the Loffoden seeds, do soak some in tepid water, and get planted with the utmost care : this is an experiment after my own heart, with chances 1000 to 1 against its success.

C. Darwin to J. D. Hooker.

Down, May 11th [1855].

My dear Hooker,—I have just received your note. I am most sincerely and heartily glad at the news * it contains, and so is my wife. Though the income is but a poor one, yet the certainty, I hope, is satisfactory to yourself and Mrs. Hooker. As it must lead in future years to the Directorship, I do hope you look at it as a piece of good fortune. For my own taste I cannot fancy a pleasanter position, than the Head of such a noble and splendid place ; far better, I should think, than a Professorship in a great town. The more I think of it, the gladder I am. But I will say no more ; except that I hope Mrs. Hooker is pretty well pleased. . . .

As the *Gardeners' Chronicle* put in my question, and took notice of it, I think I am bound to send, which I had thought of doing next week, my first report to Lindley to give him the option of inserting it ; but I think it likely that he may not think it fit for a Gardening periodical. When

---

* The appointment of Sir J. D. Hooker as Assistant Director of the Royal Gardens at Kew.

my experiments are ended (should the results appear worthy) and should the 'Linnean Journal' not object to the previous publication of imperfect and provisional reports, I should be *delighted* to insert the final report there ; for it has cost me so much trouble, that I should think that probably the result was worthy of more permanent record than a newspaper ; but I think I am bound to send it first to Lindley.

I begin to think the floating question more serious than the germinating one; and am making all the enquiries which I can on the subject, and hope to get some little light on it . . .

I hope you managed a good meeting at the Club. The Treasurership must be a plague to you, and I hope you will not be Treasurer for long : I know I would much sooner give up the Club than be its Treasurer.

Farewell, Mr. Assistant Director and dear friend,

C. DARWIN.

C. Darwin to J. D. Hooker.

June 5th, 1855.

. . . . Miss Thorley * and I are doing *a little Botanical work !* for our amusement, and it does amuse me very much, viz. making a collection of all the plants, which grow in a field, which has been allowed to run waste for fifteen years, but which before was cultivated from time immemorial ; and we are also collecting all the plants in an adjoining and *similar* but cultivated field ; just for the fun of seeing what plants have arrived or died out. Hereafter we shall want a bit of help in naming puzzlers. How dreadfully difficult it is to name plants.

What a *remarkably* nice and kind letter Dr. A. Gray has sent me in answer to my troublesome queries; I retained your copy of his 'Manual' till I heard from him, and when I have answered his letter, I will return it to you.

I thank you much for Hedysarum : I do hope it is not very

* A lady who was for many years a governess in the family.

precious, for as I told you it is for probably a *most* foolish
purpose. I read somewhere that no plant closes its leaves
so promptly in darkness, and I want to cover it up daily for
half an hour, and see if I can teach it to close by itself, or
more easily than at first in darkness. . . . . I cannot make
out why you would prefer a continental transmission, as I
think you do, to carriage by sea. I should have thought you
would have been pleased at as many means of transmission
as possible. For my own pet theoretic notions, it is quite
indifferent whether they are transmitted by sea or land, as
long as some tolerably probable way is shown. But it shocks
my philosophy to create land, without some other and independent evidence. Whenever we meet, by a very few words
I should, I think, more clearly understand your views. . . .

I have just made out my first grass, hurrah! hurrah! I
must confess that fortune favours the bold, for, as good luck
would have it, it was the easy *Anthoxanthum odoratum:*
nevertheless it is a great discovery; I never expected to
make out a grass in all my life, so hurrah! It has done my
stomach surprising good. . . .

C. Darwin to J. D. Hooker.

Down, [June?] 15th, [1855].

MY DEAR HOOKER,—I just write one line to say that the
Hedysarum is come *quite safely*, and thank you for it.

You cannot imagine what amusement you have given me
by naming those three grasses: I have just got paper to dry
and collect all grasses. If ever you catch quite a beginner,
and want to give him a taste for Botany, tell him to make
a perfect list of some little field or wood. Both Miss Thorley
and I agree that it gives a really uncommon interest to the
work, having a nice little definite world to work on, instead of
the awful abyss and immensity of all British Plants.

Adios. I was really consummately impudent to express

my opinion "on the retrograde step," * and I deserved a good snub, and upon reflection I am very glad you did not answer me in the *Gardeners' Chronicle*.

I have been *very much* interested with the Florula. †

[Writing on June 5th to Sir J. D. Hooker, my father mentions a letter from Dr. Asa Gray. The letter referred to was an answer to the following :]

C. Darwin to Asa Gray.‡

Down, April 25th [1855].

MY DEAR SIR,—I hope that you will remember that I had the pleasure of being introduced to you at Kew. I want to beg a great favour of you, for which I well know I can offer no apology. But the favour will not, I think, cause you much trouble, and will greatly oblige me. As I am no botanist, it will seem so absurd to you my asking botanical questions; that I may premise that I have for several years been collect-ing facts on "variation," and when I find that any general remark seems to hold good amongst animals, I try to test it in Plants. [Here follows a request for information on American Alpine plants, and a suggestion as to publishing on the subject.] I can assure you that I perceive how pre-sumptuous it is in me, not a botanist, to make even the most

---

* "To imagine such enormous geological changes within the period of the existence of now living beings, on no other ground but to account for their distribution, seems to me, in our present state of ignorance on the means of transportal, an almost retrograde step in science." —Extract from the paper on 'Salt Water and Seeds' in the *Gardeners' Chronicle*, May 26, 1855.

† Godron's 'Florula Juvenalis,' which gives an interesting account of

plants introduced in imported wool.

‡ The well-known American Botanist. My father's friendship with Dr. Gray began with the cor-respondence of which the present is the first letter. An extract from a letter to Sir J. Hooker, 1857, shows that my father's strong personal regard for Dr. Gray had an early origin : "I have been glad to see A. Gray's letters; there is always something in them that shows that he is a very lovable man."

trifling suggestion to such a botanist as yourself; but from what I saw and have heard of you from our dear and kind friend Hooker, I hope and think that you will forgive me, and believe me, with much respect,

Dear sir, yours very faithfully,
CHARLES DARWIN.

C. Darwin to Asa Gray.

Down, June 8th [1855].

MY DEAR SIR,—I thank you cordially for your remarkably kind letter of the 22nd ult., and for the extremely pleasant and obliging manner in which you have taken my rather troublesome questions. I can hardly tell you how much your list of Alpine plants has interested me, and I can now in some degree picture to myself the plants of your Alpine summits. The new edit. of your Manual is *capital* news for me. I know from your preface how pressed you are for room, but it would take no space to append (Eu) in brackets to any European plant, and, as far as I am concerned, this would answer every purpose.* From my own experience, whilst making out English plants in our manuals, it has often struck me how much interest it would give if some notion of their range had been given; and so, I cannot doubt, your American inquirers and beginners would much like to know which of their plants were indigenous and which European. Would it not be well in the Alpine plants to append the very same addition which you have now sent me in MS.? though here, owing to your kindness, I do not speak selfishly, but merely *pro bono Americano publico.* I presume it would be too troublesome to give in your manual the habitats of those plants found west of the Rocky Mountains, and likewise those found· in Eastern Asia, taking the Yeneseï (?),—which, if I remember right, according to Gmelin, is the main partition

---

* This suggestion Dr. Gray adopted in subsequent editions.

line of Siberia.   Perhaps Siberia more concerns the northern
Flora of North America.   The ranges of the plants to the
east and west, viz. whether most found are in Greenland and
Western Europe, or in E. Asia, appears to me a very interesting point as tending to show whether the migration has been
eastward or westward.   Pray believe me that I am most
entirely conscious that the *only use* of these remarks is to
show a botanist what points a non-botanist is curious to
learn ; for I think every one who studies profoundly a subject
often becomes unaware [on] what points the ignorant require
information.   I am so very glad that you think of drawing up
some notice on your geographical distribution, for the area
of the Manual strikes me as in some points better adapted
for comparison with Europe than that of the whole of North
America.   You ask me to state definitely some of the points
on which I much wish for information ; but I really hardly
can, for they are so vague ; and I rather wish to see what
results will come out from comparisons, than have as yet
defined objects.   I presume that, like other botanists, you
would give, for your area, the proportion (leaving out introduced plants) to the whole of the great leading families : this
is one point I had intended (and, indeed, have done roughly)
to tabulate from your book, but of course I could have done
it only *very imperfectly*.   I should also, of course, have ascertained the proportion, to the whole Flora, of the European
plants (leaving out introduced) *and of the separate great
families*, in order to speculate on means of transportal.   By
the way, I ventured to send a few days ago a copy of the
Gardeners' Chronicle with a short report by me of some
trifling experiments which I have been trying on the power
of seeds to withstand sea water.   I do not know whether
it has struck you, but it has me, that it would be advisable
for botanists to give in *whole numbers*, as well as in the
lowest fraction, the proportional numbers of the families, thus
I make out from your Manual that of the *indigenous* plants

the proportion of the Umbelliferæ are $\frac{36}{1798} = \frac{1}{49}$; for, without one knows the *whole* numbers, one cannot judge how really close the numbers of the plants of the same family are in two distant countries ; but very likely you may think this super-fluous. Mentioning these proportional numbers, I may give you an instance of the sort of points, and how vague and futile they often are, which I *attempt* to work out . . . ; reflecting on R. Brown's and Hooker's remark, that near identity of proportional numbers of the great families in two countries, shows probably that they were once continuously united, I thought I would calculate the proportions of, for instance, the *introduced* Compositæ in Great Britain to all the introduced plants, and the result was $\frac{10}{92} = \frac{1}{9.2}$. In our *aboriginal* or indigenous flora the proportion is $\frac{1}{10}$; and in many other cases I found an equally striking correspondence. I then took your Manual, and worked out the same question ; here I find in the Compositæ an almost equally striking correspondence, viz. $\frac{24}{206} = \frac{1}{8}$ in the introduced plants, and $\frac{223}{1798} = \frac{1}{8}$ in the indigenous ; but when I came to the other families I found the proportion entirely different, showing that the coincidences in the British Flora were probably accidental !

You will, I presume, give the proportion of the species to the genera, *i.e.* show on an average how many species each genus contains ; though I have done this for myself.

If it would not be too troublesome, do you not think it would be very interesting, and give a very good idea of your Flora, to divide the species into three groups, viz. (*a*) species com-mon to the old world, stating numbers common to Europe and Asia ; (*b*) indigenous species, but belonging to genera found in the Old World ; and (*c*) species belonging to genera confined to America or the New World ? To make (according to my ideas) perfection perfect, one ought to be told whether there are other cases, like Erica, of genera common in Europe or in Old World not found in your area. But honestly I feel

that it is quite ridiculous my writing to you at such length on
the subject ; but, as you have asked me, I do it gratefully, and
write to you as I should to Hooker, who often laughs at me
unmercifully, and I am sure you have better reason to do so.

There is one point on which I am *most* anxious for information, and I mention it with the greatest hesitation, and
only in the *full belief* that you will believe me that I have
not the folly and presumption to hope for a second that you
will give it, without you can with very little trouble. The
point can at present interest no one but myself, which makes
the case wholly different from geographical distribution. The
only way in which, I think, you possibly could do it with little
trouble would be to bear in mind, whilst correcting your proof-
sheets of the Manual, my question and put a cross or mark
to the species, and whenever sending a parcel to Hooker to
let me have such old sheets. But this would give you the
trouble of remembering my question, and I can hardly hope
or expect that you will do it. But I will just mention what I
want ; it is to have marked the " close species " in a Flora, so
as to compare in *different* Floras whether the same genera
have "close species," and for other purposes too vague to
enumerate. I have attempted, by Hooker's help, to ascertain
in a similar way whether the different species of the same
genera in distant quarters of the globe are variable or
present varieties. The definition I should give of a "*close
species*" was one that *you* thought specifically distinct, but
which you could conceive some other *good* botanist might
think only a race or variety ; or, again, a species that you
had trouble, though having opportunities of knowing it well,
in discriminating from some other species. Supposing that
you were inclined to be so very kind as to do this, and could
(which I do not expect) spare the time, as I have said, a
mere cross to each such species in any useless proof-sheets
would give me the information desired, which, I may add,
I know must be vague.

How can I apologise enough for all my presumption and the extreme length of this letter? The great good nature of your letter to me has been partly the cause, so that, as is too often the case in this world, you are punished for your good deeds. With hearty thanks, believe me,

Yours very truly and gratefully,

CH. DARWIN.

C. Darwin to J. D. Hooker.

Down, 18th [July, 1855].

. . . I think I am getting a *mild* case about Charlock seed;* but just as about salting, ill luck to it, I cannot remember how many years you would allow that Charlock seed might live in the ground. Next time you write, show a bold face, and say in how many years, you think, Charlock seed would probably all be dead. A man told me the other day of, as I thought, a splendid instance,—and *splendid* it was, for according to his evidence the seed came up alive out of the *lower part* of the *London Clay ! ! !* I disgusted him by telling him that Palms ought to have come up.

You ask how far I go in attributing organisms to a common descent: I answer I know not; the way in which I intend treating the subject, is to show (*as far as I can*) the facts and arguments for and against the common descent of the species of the same genus; and then show how far the same arguments tell for or against forms, more and more widely different: and when we come to forms of different orders and

---

* In the *Gardeners' Chronicle*, 1855, p. 758, appeared a notice (half a column in length) by my father on the "Vitality of Seeds." The facts related refer to the "Sand-walk"; the wood was planted in 1846 on a piece of pasture land laid down as grass in 1840. In 1855, on the soil being dug in several places, Charlock (*Brassica sinapistrum*) sprang up freely. The subject continued to interest him, and I find a note dated July 2nd, 1874, in which my father recorded that forty-six plants of Charlock sprang up in that year over a space (14 × 7 feet) which had been dug to a considerable depth.

classes, there remain only some such arguments as those which can perhaps be deduced from similar rudimentary structures, and very soon not an argument is left.

[The following extract from a letter to Mr. Fox [Oct. 1855* gives a brief mention of the last meeting of the British Association which he attended :] "I really have no news: the only thing we have done for a long time, was to go to Glasgow ; but the fatigue was to me more than it was worth, and E. caught a bad cold. On our return we stayed a single day at Shrewsbury, and enjoyed seeing the old place. I saw a little of Sir Philip † (whom I liked much), and he asked me 'why on earth I instigated you to rob his poultry-yard ?' The meeting was a good one, and the Duke of Argyll spoke excellently."]

* In this year he published ('Phil. Mag.' x.) a paper " On the power of icebergs to make rectilinear uniformly-directed grooves across a submarine undulatory surface."

† Sir P. Egerton was a neighbour of Mr. Fox.

CHAPTER III.

THE UNFINISHED BOOK.

MAY 1856 TO JUNE 1858.

[IN the Autobiographical chapter (Vol. I. p. 84) my father wrote:—"Early in 1856 Lyell advised me to write out my views pretty fully, and I began at once to do so on a scale three or four times as extensive as that which was afterwards followed in my 'Origin of Species;' yet it was only an abstract of the materials which I had collected." The letters in the present chapter are chiefly concerned with the preparation of this unfinished book.

The work was begun on May 14th, and steadily continued up to June 1858, when it was interrupted by the arrival of Mr. Wallace's MS. During the two years which we are now considering, he wrote ten chapters (that is about one-half) of the projected book. He remained for the most part at home, but paid several visits to Dr. Lane's Water-Cure Establishment at Moor Park, during one of which he made a pilgrimage to the shrine of Gilbert White at Selborne.]

LETTERS.

C. Darwin to C. Lyell.

May 3 [1856].

. . . With respect to your suggestion of a sketch of my views, I hardly know what to think, but will reflect on it, but

F 2

it goes against my prejudices. To give a fair sketch would be absolutely impossible, for every proposition requires such an array of facts.   If I were to do anything, it could only refer to the main agency of change—selection—and perhaps point out a very few of the leading features, which countenance such a view, and some few of the main difficulties.   But I do not know what to think ; I rather hate the idea of writing for priority, yet I certainly should be vexed if any one were to publish my doctrines before me.   Anyhow, I thank you heartily for your sympathy.   I shall be in London next week, and I will call on you on Thursday morning for one hour precisely, so as not to lose much of your time and my own ; but will you let me this time come as early as 9 o'clock, for I have much which I must do in the morning in my strongest time ?   Farewell, my dear old patron.

Yours,
C. DARWIN.

By the way, *three* plants have come up out of the earth, perfectly enclosed in the roots of the trees. And twenty-nine plants in the table-spoonful of mud, out of the little pond ; Hooker was surprised at this, and struck with it, when I showed him how much mud I had scraped off one duck's feet.

If I did publish a short sketch, where on earth should I publish it ?

If I do *not* hear, I shall understand that I may come from 9 to 10 on Thursday.

C. Darwin to J. D. Hooker.

May 9th [1856].

. . . I very much want advice and *truthful* consolation if you can give it. I had a good talk with Lyell about my species work, and he urges me strongly to publish something. I am fixed against any periodical or Journal, as I positively will *not* expose myself to an Editor or a Council, allowing a publication for which they might be abused. If I publish anything it must be a *very thin* and little volume, giving a sketch of my views and difficulties ; but it is really dreadfully unphilosophical to give a *résumé*, without exact references, of an unpublished work. But Lyell seemed to think I might do this, at the suggestion of friends, and on the ground, which I might state, that I had been at work for eighteen * years, and yet could not publish for several years, and especially as I could point out difficulties which seemed to me to require especial investigation. Now what think you ? I should be really grateful for advice. I thought of giving up a couple of months and writing such a sketch, and trying to keep my judgment open whether or no to publish it when completed. It will be simply impossible for me to give exact references ; anything important I should state on the authority of the author generally ; and instead of giving all the facts on which I ground my opinion, I could give by memory only one or two. In the Preface I would state that the work could not be considered strictly scientific, but a mere sketch or outline of a future work in which full references, &c., should be given. Eheu, eheu, I believe I should sneer at any one else doing this, and my only comfort is, that I *truly* never dreamed of it, till Lyell suggested it, and seems deliberately to think it advisable.

I am in a peck of troubles, and do pray forgive me for troubling you.

Yours affectionately,
C. DARWIN.

C. Darwin to J. D. Hooker.

May 11th [1856].

. . . Now for a *more important !* subject, viz. my own self : I am extremely glad you think well of a separate "Pre-

---

* The interval of eighteen years, from 1837 when he began to collect facts, would bring the date of this letter to 1855, not 1856, nevertheless the latter seems the more probable date.

liminary Essay " (*i.e.* if anything whatever is published; for Lyell seemed rather to doubt on this head)* ; but I cannot bear the idea of *begging* some Editor and Council to publish, and then perhaps to have to *apologise* humbly for having led them into a scrape.   In this one respect I am in the state which, according to a very wise saying of my father's, is the only fit state for asking advice, viz. with my mind firmly made up, and then, as my father used to say, *good* advice was very comfortable, and it was easy to reject *bad* advice.   But Heaven knows I am not in this state with respect to publishing at all any preliminary essay.   It yet strikes me as quite unphilosophical to publish results without the full details which have led to such results.

It is a melancholy, and I hope not quite true view of yours that facts will prove anything, and are therefore superfluous! But I have rather exaggerated, I see, your doctrine.   I do not fear being tied down to error, *i.e.* I feel pretty sure I should give up anything false published in the preliminary essay, in my larger work ; but I may thus, it is very true, do mischief by spreading error, which as I have often heard you say is much easier spread than corrected.   I confess I lean more and more to at least making the attempt and drawing up a sketch and trying to keep my judgment, whether to publish, open.   But I always return to my fixed idea that it is dreadfully unphilosophical to publish without full details. I certainly think my future work in full would profit by hearing what my friends or critics (if reviewed) thought of the outline.

To any one but you I should apologise for such long discus= sion on so personal an affair ; but I believe, and indeed you have proved it by the trouble you have taken, that this would be superfluous.

Yours truly obliged,

CH. DARWIN.

* The meaning of the sentence in parentheses is obscure.

P.S.—What you say (for I have just re-read your letter) that the Essay might supersede and take away all novelty and value from any future larger Book, is very true; and that would grieve me beyond everything. On the other hand (again from Lyell's urgent advice), I published a preliminary sketch of the Coral Theory, and this did neither good nor harm. I begin *most heartily* to wish that Lyell had never put this idea of an Essay into my head.

From a Letter to Sir C. Lyell [*July*, 1856].

"I am delighted that I may say (with absolute truth) that my essay is published at your suggestion, but I hope it will not need so much apology as I at first thought; for I have resolved to make it nearly as complete as my present materials allow. I cannot put in all which you suggest, for it would appear too conceited."

From a Letter to W. D. Fox.

Down, June 14th [1856].

". . . What you say about my Essay, I dare say is very true; and it gave me another fit of the wibber-gibbers: I hope that I shall succeed in making it modest. One great motive is to get information on the many points on which I want it. But I tremble about it, which I should not do, if I allowed some three or four more years to elapse before publishing anything. . . ."

[The following extracts from letters to Mr. Fox are worth giving, as showing how great was the accumulation of material which now had to be dealt with.

June 14th [1856].

"Very many thanks for the capital information on cats; I see I had blundered greatly, but I know I have somewhere your original notes; but my notes are so numerous during nineteen years' collection, that it would take me at least a year to go over and classify them."

Nov. 1856. "Sometimes I fear I shall break down, for my subject gets bigger and bigger with each month's work."]

C. Darwin to C. Lyell.

Down, 16th [June, 1856].

MY DEAR LYELL,—I am going to do the most impudent thing in the world. But my blood gets hot with passion and turns cold alternately at the geological strides, which many of your disciples are taking.

Here, poor Forbes made a continent to [*i.e.* extending to] North America and another (or the same) to the Gulf weed; Hooker makes one from New Zealand to South America and round the World to Kerguelen Land. Here is Wollaston speaking of Madeira and P. Santo "as the sure and certain witnesses of a former continent." Here is Woodward writes to me, if you grant a continent over 200 or 300 miles of ocean depths (as if that was nothing), why not extend a continent to every island in the Pacific and Atlantic Oceans? And all this within the existence of recent species! If you do not stop this, if there be a lower region for the punishment of geologists, I believe, my great master, you will go there. Why, your disciples in a slow and creeping manner beat all the old Catastrophists who ever lived. You will live to be the great chief of the Catastrophists.

There, I have done myself a great deal of good, and have exploded my passion.

So my master, forgive me, and believe me, ever yours,

C. DARWIN.

P.S.—Don't answer this, I did it to ease myself.

C. Darwin to J. D. Hooker.

Down [June] 17th, 1856.

. . . I have been very deeply interested by Wollaston's book,* though I differ *greatly* from many of his doctrines. Did you ever read anything so rich, considering how very far he goes, as his denunciations against those who go further : " most mischievous," " absurd," " unsound." Theology is at the bottom of some of this. I told him he was like Calvin burning a heretic. It is a very valuable and clever book in my opinion. He has evidently read very little out of his own line. I urged him to read the New Zealand essay. His Geology also is rather *eocene*, as I told him. In fact I wrote most frankly ; I fear too frankly ; he says he is sure that ultra-honesty is my characteristic : I do not know whether he meant it as a sneer ; I hope not. Talking of eocene geology, I got so wroth about the Atlantic continent, more especially from a note from Woodward (who has published a capital book on shells), who does not seem to doubt that every island in the Pacific and Atlantic are the remains of continents, submerged within period of existing species, that I fairly exploded, and wrote to Lyell to protest, and summed up all the continents created of late years by Forbes (the head sinner !) *yourself*, Wollaston, and Woodward, and a pretty nice little extension of land they make altogether ! I am fairly rabid on the question and therefore, if not wrong already, am pretty sure to become so . . .

I have enjoyed your note much. Adios,

C. DARWIN.

P.S. [June] 18th.—Lyell has written me a *capital* letter on your side, which ought to upset me entirely, but I cannot say it does quite.

Though I must try and cease being rabid and try to feel

* ' The Variation of Species,' 1856.

humble, and allow you all to make continents, as easily as a cook does pancakes.

C. Darwin to C. Lyell.

Down, June 25th [1856].

MY DEAR LYELL,—I will have the following tremendous letter copied to make the reading easier, and as I want to keep a copy.

As you say you would like to hear my reasons for being most unwilling to believe in the continental extensions of late authors, I gladly write them, as, without I am convinced of my error, I shall have to give them condensed in my essay, when I discuss single and multiple creation ; I shall therefore be particularly glad to have your general opinion on them. I may *quite likely* have persuaded myself in my wrath that there is more in them than there is.   If there was much more reason to admit a continental extension in any one or two instances (as in Madeira) than in other cases, I should feel no difficulty whatever.   But if on account of European plants, and littoral sea shells, it is thought necessary to join Madeira to the mainland, Hooker is quite right to join New Holland to New Zealand, and Auckland Island (and Raoul Island to N.E.), and these to S. America and the Falklands, and these to Tristan d'Acunha, and these to Kerguelen Land ; thus making, either strictly at the same time, or at different periods, but all within the life of recent beings, an almost circumpolar belt of land.   So again Galapagos and Juan Fernandez must be joined to America ; and if we trust to littoral sea shells, the Galapagos must have been joined to the Pacific Islands (2400 miles distant) as well as to America, and as Woodward seems to think all the islands in the Pacific into a magnificent con-tinent ; also the islands in the Southern Indian Ocean into another continent, with Madagascar and Africa, and perhaps India.   In the North Atlantic, Europe will stretch half-way

across the ocean to the Azores, and further north right across.
In short, we must suppose probably, half the present ocean
was land within the period of living organisms. The Globe
within this period must have had a quite different aspect.
Now the only way to test this, that I can see, is to consider
whether the continents have undergone within this same period such wonderful permutations. In all North and South
and Central America, we have both recent and miocene (or
eocene) shells, quite distinct on the opposite sides, and hence
I cannot doubt that *fundamentally* America has held its place
since at least, the miocene period. In Africa almost all the
living shells are distinct on the opposite sides of the inter-
tropical regions, short as the distance is compared to the range
of marine mollusca, in uninterrupted seas ; hence I infer that
Africa has existed since our present species were created.
Even the isthmus of Suez and the Aralo-Caspian basin have
had a great antiquity. So I imagine, from the tertiary deposits, has India. In Australia the great fauna of extinct marsupials shows that before the present mammals appeared,
Australia was a separate continent. I do not for one second
doubt that very large portions of all these continents have
undergone *great* changes of level within this period, but yet I
conclude that fundamentally they stood as barriers in the sea,
where they now stand ; and therefore I should require the
weightiest evidence to make me believe in such immense
changes within the period of living organisms in our oceans,
where, moreover, from the great depths, the changes must
have been vaster in a vertical sense.

Secondly. Submerge our present continents, leaving a few
mountain peaks as islands, and what will the character of the
islands be ?—Consider that the Pyrenees, Sierra Nevada,
Apennines, Alps, Carpathians, are non-volcanic, Etna and
Caucasus, volcanic. In Asia, Altai and Himalaya, I believe
non-volcanic. In North Africa the non-volcanic, as I imagine,
Alps of Abyssinia and of the Atlas. In South Africa, the

Snow Mountains. In Australia, the non-volcanic Alps. In North America, the White Mountains, Alleghanies and Rocky Mountains—some of the latter alone, I believe, volcanic. In South America to the east, the non-volcanic [Silla] of Caracas, and Itacolumi of Brazil, further south the Sierra |Ventanas, and in the Cordilleras, many volcanic but not all. Now compare these peaks with the oceanic islands; as far as known all are volcanic, except St. Paul's (a strange bedevilled rock), and the Seychelles, if this latter can be called oceanic, in the line of Madagascar; the Falklands, only 500 miles off, are only a shallow bank; New Caledonia, hardly oceanic, is another exception. This argument has to me great weight. Compare on a Geographical Map, islands which, we have *several* reasons to suppose, were connected with mainland, as Sardinia, and how different it appears. Believing, as I am inclined, that continents as continents, and oceans as oceans, are of immense antiquity—I should say that if any of the existing oceanic islands have any relation of any kind to continents, they are forming continents; and that by the time they could form a continent, the volcanoes would be denuded to their cores, leaving peaks of syenite, diorite, or porphyry. But have we nowhere any last wreck of a continent, in the midst of the ocean? St. Paul's Rock, and such old battered volcanic islands, as St. Helena, may be; but I think we can see some reason why we should have less evidence of sinking than of rising continents (if my view in my Coral volume has any truth in it, viz.: that volcanic outbursts accompany rising areas), for during subsidence there will be no compensating agent at work, in rising areas there will be the *additional* element of outpoured volcanic matter.

Thirdly. Considering the depth of the ocean, I was, before I got your letter, inclined vehemently to dispute the vast amount of subsidence, but I must strike my colours. With respect to coral reefs, I carefully guarded against its being supposed that a continent was indicated by the groups of atolls. It is difficult to guess, as it seems to me, the amount of subsidence indicated by coral reefs; but in such large areas as the Lowe Archipelago, the Marshall Archipelago, and Laccadive group, it would, judging from the heights of existing oceanic archipelagoes, be odd, if some peaks of from 8000 to 10,000 feet had not been buried. Even after your letter a suspicion crossed me whether it would be fair to argue from subsidences in the middle of the greatest oceans to continents; but refreshing my memory by talking with Ramsay in regard to the probable thickness in one vertical line of the Silurian and carboniferous formation, it seems there must have been *at least* 10,000 feet of subsidence during these formations in Europe and North America, and therefore during the continuance of nearly the same set of organic beings. But even 12,000 feet would not be enough for the Azores, or for Hooker's continent; I believe Hooker does not infer a continuous continent, but approximate groups of islands, with, if we may judge from existing continents, not *profoundly* deep sea between them; but the argument from the volcanic nature of nearly every existing oceanic island tells against such supposed groups of islands,—for I presume he does not suppose a mere chain of volcanic islands belting the southern hemisphere.

Fourthly. The supposed continental extensions do not seem to me, perfectly to account for all the phenomena of distribution on islands; as the absence of mammals and Batrachians; the absence of certain great groups of insects on Madeira, and of Acaciæ and Banksias, &c., in New Zealand; the paucity of plants in some cases, &c. Not that those who believe in various accidental means of dispersal, can explain most of these cases; but they may at least say that these facts seem hardly compatible with former continuous land.

Finally. For these several reasons, and especially considering it certain (in which you will agree) that we are extremely ignorant of means of dispersal, I cannot avoid think-ing that Forbes' 'Atlantis' was an ill-service to science, as checking a close study of means of dissemination. I shall be really grateful to hear, as briefly as you like, whether these arguments have any weight with you, putting yourself in the position of an honest judge. I told Hooker I was going to write to you on this subject; and I should like him to read this ; but whether he or you will think it worth time and postage remains to be proved.

Yours most truly,

CHARLES DARWIN.

[On July 8th he wrote to Sir Charles Lyell.

"I am sorry you cannot give any verdict on Continental extensions; and I infer that you think my argument of not much weight against such extensions. I know I wish I could believe so."]

C. Darwin to Asa Gray.

Down, July 20th [1856].

. . . It is not a little egotistical, but I should like to tell you (and I do not *think* I have) how I view my work. Nineteen years (!) ago it occurred to me that whilst otherwise employed on Nat. Hist., I might perhaps do good if I noted any sort of facts bearing on the question of the origin of species, and this I have since been doing. Either species have been independently created, or they have descended from other species, like varieties from one species. I think it can be shown to be probable that man gets his most distinct varieties by preserving such as arise best worth keeping and destroying the others, but I should fill a quire if I were to go on. To be brief, I *assume* that species arise like our domestic varieties with *much* extinction ; and then test this hypothesis by comparison with as many general and pretty well-esta-blished propositions as I can find made out,—in geographical

distribution, geological history, affinities, &c. &c. And it
seems to me that, *supposing* that such hypothesis were to
explain such general propositions, we ought, in accordance
with the common way of following all sciences, to admit it till
some better hypothesis be found out. For to my mind to
say that species were created so and so is no scientific explanation, only a reverent way of saying it is so and so. But it
is nonsensical trying to show how I try to proceed, in the
compass of a note. But as an honest man, I must tell you that
I have come to the heterodox conclusion, that there are no
such things as independently created species—that species are
only strongly defined varieties. I know that this will make
you despise me. I do not much underrate the many *huge*
difficulties on this view, but yet it seems to me to explain too
much, otherwise inexplicable, to be false. Just to allude to
one point in your last note, viz. about species of the same
genus *generally* having a common or continuous area ; if they
are actual lineal descendants of one species, this of course
would be the case ; and the sadly too many exceptions (for
me) have to be explained by climatal and geological changes.
A fortiori on this view (but on exactly same grounds), all the
individuals of the same species should have a continuous
distribution. On this latter branch of the subject I have put
a chapter together, and Hooker kindly read it over. I
thought the exceptions and difficulties were so great that on
the whole the balance weighed against my notions, but I was
much pleased to find that it seemed to have considerable
weight with Hooker, who said he had never been so much
staggered about the permanence of species.

I must say one word more in justification (for I feel sure
that your tendency will be to despise me and my crotchets),
that all my notions about *how* species change are derived
from long-continued study of the works of (and converse
with) agriculturists and horticulturists ; and I believe I
see my way pretty clearly on the means used by nature to

change her species and *adapt* them to the wondrous and exquisitely beautiful contingencies to which every living being
is exposed. . . .

C. Darwin to J. D. Hooker.

Down, July 30th, 1856.

MY DEAR HOOKER,—Your letter is of *much* value to me.
I was not able to get a definite answer from Lyell,* as you will
see in the enclosed letters, though I inferred that he thought
nothing of my arguments. Had it not been for this correspondence, I should have written sadly too strongly. You
may rely on it I shall put my doubts moderately. There
never was such a predicament as mine : here you continental
extensionists would remove enormous difficulties opposed to
me, and yet I cannot honestly admit the doctrine, and must
therefore say so. I cannot get over the fact that not a fragment of secondary or palæozoic rock has been found on any
island above 500 or 600 miles from a mainland. You rather
misunderstand me when you think I doubt the *possibility* of
subsidence of 20,000 or 30,000 feet ; it is only probability, considering such evidence as we have independently of distribution.
I have not yet worked out in full detail the distribution of
mammalia, both *identical* and allied, with respect to the *one
element of depth of the sea*; but as far as I have gone, the
results are to me surprisingly accordant with my very most
troublesome belief in not such great geographical changes as
you believe ; and in mammalia we certainly know more of
means of distribution than in any other class. Nothing is so
vexatious to me, as so constantly finding myself drawing
different conclusions from better judges than myself, from
the same facts.

I fancy I have lately removed many (not geographical)
great difficulties opposed to my notions, but God knows it
may be all hallucination.

* On the continental extensions of Forbes and others.

Please return Lyell's letters.

What a capital letter of Lyell's that to you is, and what a wonderful man he is. I differ from him greatly in thinking that those who believe that species are *not* fixed will multiply specific names : I know in my own case my most frequent source of doubt was whether others would not think this or that was a God-created Barnacle, and surely deserved a name. Otherwise I should only have thought whether the amount of difference and permanence was sufficient to justify a name : I am, also, surprised at his thinking it immaterial whether species are absolute or not : whenever it is proved that all species are produced by generation, by laws of change, what good evidence we shall have of the gaps in formations. And what a science Natural History will be, when we are in our graves, when all the laws of change are thought one of the most important parts of Natural History.

I cannot conceive why Lyell thinks such notions as mine or of 'Vestiges,' will invalidate specific centres. But I must not run on and take up your time. My MS. will not, I fear, be copied before you go abroad. With hearty thanks.

Ever yours,

C. DARWIN.

P.S.—After giving much condensed, my argument versus continental extensions, I shall append some such sentence, as that two better judges than myself have considered these arguments, and attach no weight to them.

C. Darwin to J. D. Hooker.

Down, August 5th [1856].

. . . I quite agree about Lyell's letters to me, which, though to me interesting, have afforded me no new light. Your letters, under the *geological* point of view, have been more valuable to me. You cannot imagine how earnestly I wish I could swallow continental extension, but I cannot ;

the more I think (and I cannot get the subject out of my
head), the more difficult I find it. If there were only some
half-dozen cases, I should not feel the least difficulty ; but
the generality of the facts of all islands (except one or two)
having a considerable part of their productions in common
with one or more mainlands utterly staggers me. What a
wonderful case of the Epacridæ ! It is most vexatious, also
humiliating, to me that I cannot follow and subscribe to the
way in which you strikingly put your view of the case.
I look at your facts (about Eucalyptus, &c.) as *damning*
against continental extension, and if you like also damning
against migration, or at least of *enormous* difficulty. I see
the ground of our difference (in a letter I must put myself
on an equality in arguing) lies, in my opinion, that scarcely
anything is known of means of distribution. I quite agree
with A. De Candolle's (and I dare say your) opinion that it
is poor work putting together the merely *possible* means of
distribution ; but I see no other way in which the subject can
be attacked, for I think that A. De Candolle's argument,
that no plants have been introduced into England except by
man's agency, of no weight. I cannot but think that the
theory of continental extension does do some little harm
as stopping investigation of the means of dispersal, which,
whether *negative* or positive, seems to me of value ; when
negatived, then every one who believes in single centres will
have to admit continental extensions.

. . . I see from your remarks that you do not understand
my notions (whether or no worth anything) about modification ; I attribute very little to the direct action of climate, &c.
I suppose, in regard to specific centres, we are at cross
purposes ; I should call the kitchen garden in which the red
cabbage was produced, or the farm in which Bakewell made
the Shorthorn cattle, the specific centre of these *species !*
And surely this is centralisation enough !

I thank you most sincerely for all your assistance ; and

whether or no my book may be wretched, you have done your best to make it less wretched. Sometimes I am in very good spirits and sometimes very low about it. My own mind is decided on the question of the origin of species ; but, good heavens, how little that is worth ! . . .

[With regard to " specific centres," a passage from a letter dated July 25, 1856, from Sir Charles Lyell to Sir J. D. Hooker (' Life,' vol. ii. p. 216) is of interest :

" I fear much that if Darwin argues that species are phantoms, he will also have to admit that single centres of dispersion are phantoms also, and that would deprive me of much of the value which I ascribe to the present provinces of animals and plants, as illustrating modern and tertiary changes in physical geography."

He seems to have recognised, however, that the phantom doctrine would soon have to be faced, for he wrote in the same letter : " Whether Darwin persuades you and me to renounce our faith in species (when geological epochs are considered) or not, I foresee that many will go over to the indefinite modifiability doctrine."

In the autumn my father was still working at geographical distribution, and again sought aid from Sir J. D. Hooker.

" In the course of some weeks, you unfortunate wretch, you will have my MS. on one point of Geographical Distribution. I will, however, never ask such a favour again ; but in regard to this one piece of MS., it is of infinite importance to me for you to see it ; for never in my life have I felt such difficulty what to do, and I heartily wish I could slur the whole subject over."

In a letter to Sir J. D. Hooker (June, 1856), the following characteristic passage occurs, suggested, no doubt, by the kind of work which his chapter on Geographical Distribution entailed :

"There is wonderful ill logic in his [E. Forbes'] famous and admirable memoir on distribution, as it appears to me, now that I have got it up so as to give the heads in a page. Depend on it, my saying is a true one, viz. that a compiler is a *great* man, and an original man a commonplace man. Any fool can generalise and speculate; but, oh, my heavens! to get up *at second hand* a New Zealand Flora, that is work."]

C. Darwin to W. D. Fox.

Oct. 3 [1856].

. . . I remember you protested against Lyell's advice of writing a *sketch* of my species doctrines. Well, when I began I found it such unsatisfactory work that I have desisted, and am now drawing up my work as perfect as my materials of nineteen years' collecting suffice, but do not intend to stop to perfect any line of investigation beyond current work. Thus far and no farther I shall follow Lyell's urgent advice. Your remarks weighed with me considerably. I find to my sorrow it will run to quite a big book. I have found my careful work at pigeons really invaluable, as enlightening me on many points on variation under domestication. The copious old literature, by which I can trace the gradual changes in the breeds of pigeons has been extraordinarily useful to me. I have just had pigeons and fowls *alive* from the Gambia ! Rabbits and ducks I am attending to pretty carefully, but less so than pigeons. I find most remarkable differences in the skeletons of rabbits. Have you ever kept any odd breeds of rabbits, and can you give me any details ? One other question. You used to keep hawks; do you at all know, after eating a bird, how soon after they throw up the pellet ?

No subject gives me so much trouble and doubt and difficulty as the means of dispersal of the same species of terrestrial
productions on the oceanic islands.　Land mollusca drive me
mad, and I cannot anyhow get their eggs to experimentise their
power of floating and resistance to the injurious action of
salt water.　I will not apologise for writing so much about
my own doings, as I believe you will like to hear.　Do some-
time, I beg you, let me hear how you get on in health ; and
if so inclined, let me have some words on call-ducks.

My dear Fox, yours affectionately,

CH. DARWIN.

[With regard to his book he wrote (Nov. 10th) to Sir
Charles Lyell :

"I am working very steadily at my big book ; I have
found it quite impossible to publish any preliminary essay or
sketch ; but am doing my work as completely as my present
materials allow without waiting to perfect them.　And this
much acceleration I owe to you."]

C. Darwin to J. D. Hooker.

Down, Sunday [Oct. 1856].

MY DEAR HOOKER,—The seeds are come all safe, many
thanks for them.　I was very sorry to run away so soon and
miss any part of my *most* pleasant evening ; and I ran away
like a Goth and Vandal without wishing Mrs. Hooker good-
bye ; but I was only just in time, as I got on the platform
the train had arrived.

I was particularly glad of our discussion after dinner ;
fighting a battle with you always clears my mind wonderfully.　I groan to hear that A. Gray agrees with you about
the condition of Botanical Geography.　All I know is that
if you had had to search for light in Zoological Geography
you would by contrast, respect your own subject a vast deal

more than you now do. The hawks have behaved like gentlemen, and have cast up pellets with lots of seeds in them ; and I have just had a parcel of partridge's feet well caked with mud ! ! ! *   Adios.

Your insane and perverse friend,

C. DARWIN.

C. Darwin to J. D. Hooker.

Down, Nov. 4th [1856].

MY DEAR HOOKER,—I thank you more *cordially* than you will think probable, for your note. Your verdict † has been a great relief. On my honour I had no idea whether or not you would say it was (and I knew you would say it very kindly) so bad, that you would have begged me to have burnt the whole. To my own mind my MS. relieved me of some few difficulties, and the difficulties seemed to me pretty fairly stated, but I had become so bewildered with conflicting facts, evidence, reasoning and opinions, that I felt to myself that I had lost all judgment. Your general verdict is *incomparably* more favourable than I had anticipated . . .

C. Darwin to J. D. Hooker.

Down, Nov. 23rd [1856].

MY DEAR HOOKER,—I fear I shall weary you with letters, but do not answer this, for in truth and without flattery, I so value your letters, that after a heavy batch, as of late, I feel that I have been extravagant and have drawn too much money, and shall therefore have to stint myself on another occasion.

When I sent my MS. I felt strongly that some preliminary questions on the causes of variation ought to have been sent you. Whether I am right or wrong in these points is quite a

---

* The mud in such cases often contains seeds, so that plants are thus transported.

† On the MS. relating to geographical distribution.

separate question, but the conclusion which I have come to,
quite independently of geographical distribution, is that
external conditions (to which naturalists so often appeal) do
by themselves *very little*. How much they do is the point of
all others on which I feel myself very weak. I judge from
the facts of variation under domestication, and I may yet get
more light. But at present, after drawing up a rough copy
on this subject, my conclusion is that external conditions do
extremely little, except in causing mere variability. This
mere variability (causing the child *not* closely to resemble its
parent) I look at as *very* different from the formation of a
marked variety or new species. (No doubt the variability is
governed by laws, some of which I am endeavouring very
obscurely to trace.) The formation of a strong variety or
species I look at as almost wholly due to the selection of
what may be incorrectly called *chance* variations or variability.
This power of selection stands in the most direct relation to
time, and in the state of nature can be only excessively slow.
Again, the slight differences selected, by which a race or
species is at last formed, stands, as I think can be shown
(even with plants, and obviously with animals), in a far more
important relation to its associates than to external conditions.
Therefore, according to my principles, whether right or wrong,
I cannot agree with your proposition that time, and altered
conditions, and altered associates, are " convertible terms." I
look at the first and the last as *far* more important : time
being important only so far as giving scope to selection.
God knows whether you will perceive at what I am driving.
I shall have to discuss and think more about your difficulty of
the temperate and sub-arctic forms in the S. hemisphere than
I have yet done. But I am inclined to think that I am right
(if my general principles are right), that there would be little
tendency to the formation of a new species, during the period
of migration, whether shorter or longer, though considerable
variability may have supervened. . . .

C. Darwin to J. D. Hooker.

Dec. 24th [1856].

. . . How I do wish I lived near you to discuss matters with. I have just been comparing definitions of species, and stating briefly how systematic naturalists work out their subjects. Aquilegia in the Flora Indica was a capital example for me. It is really laughable to see what different ideas are prominent in various naturalists' minds, when they speak of "species;" in some, resemblance is everything and descent of little weight—in some, resemblance seems to go for nothing, and Creation the reigning idea—in some, descent is the key,—in some, sterility an unfailing test, with others it is not worth a farthing. It all comes, I believe, from trying to define the undefinable. I suppose you have lost the odd black seed from the birds' dung, which germinated,—anyhow, it is not worth taking trouble over. I have now got about a dozen seeds out of small birds' dung. Adios,

My dear Hooker, ever yours,

C. DARWIN.

C. Darwin to Asa Gray.

Down, Jan. 1st [1857 ?]

MY DEAR DR. GRAY,—I have received the second part of your paper,* and though I have nothing particular to say, I must send you my thanks and hearty admiration. The whole paper strikes me as quite exhausting the subject, and I quite fancy and flatter myself I now appreciate the character of your Flora. What a difference in regard to Europe your remark in relation to the genera makes! I have been eminently glad to see your conclusion in regard to the species of large genera widely ranging; it is in strict conformity with

* 'Statistics of the Flora of the Northern United States.'—*Silliman's Journal,* 1857.

the results I have worked out in several ways.   It is of great
importance to my notions.   By the way you have paid me a
great compliment : * to be *simply* mentioned even in such a
paper I consider a very great honour.   One of your conclusions makes me groan, viz. that the line of connection of
the strictly Alpine plants is through Greenland.   I should
extremely like to see your reasons published in detail, for it
"riles" me (this is a proper expression, is it not?) dreadfully.
Lyell told me, that Agassiz having a theory about when
Saurians were first created, on hearing some careful observations opposed to this, said he did not believe it, " for Nature
never lied."   I am just in this predicament, and repeat to
you that, " Nature never lies," ergo, theorisers are always
right. . . .

Overworked as you are, I dare say you will say that I am
an odious plague ; but here is another suggestion !   I was led
by one of my wild speculations to conclude (though it has
nothing to do with geographical distribution, yet it has with
your statistics) that trees would have a strong tendency to have
flowers with diœcious, monœcious or polygamous structure.
Seeing that this seemed so in Persoon, I took one little
British Flora, and discriminating trees from bushes according
to Loudon, I have found that the result was in species, genera
and families, as I anticipated.   So I sent my notions to Hooker
to ask him to tabulate the New Zealand Flora for this end,
and he thought my result sufficiently curious, to do so ; and
the accordance with Britain is very striking, and the more so,
as he made three classes of trees, bushes, and herbaceous
plants.   (He says further he shall work the Tasmanian Flora
on the same principle.)   The bushes hold an intermediate
position between the other two classes.   It seems to me a

* " From some investigations of   range over a larger area than the
his own, this sagacious naturalist   species of small genera do."—Asa
inclines to think that large genera   Gray, *loc. cit.*

curious relation in itself, and is very much so, if my theory and explanation are correct.*

With hearty thanks, your most troublesome friend,

C. DARWIN.

C. Darwin to J. D. Hooker.

Down, April 12th [1857].

MY DEAR HOOKER,—Your letter has pleased me much, for I never can get it out of my head, that I take unfair advantage of your kindness, as I receive all and give nothing. What a splendid discussion you could write on the whole subject of variation! The cases discussed in your last note are valuable to me (though odious and damnable), as showing how profoundly ignorant we are on the causes of variation. I shall just allude to these cases, as a sort of sub-division of polymorphism a little more definite, I fancy, than the variation of, for instance, the Rubi, and equally or more perplexing.

I have just been putting my notes together on variations *apparently* due to the immediate and direct action of external causes; and I have been struck with one result. The most firm sticklers for independent creation admit, that the fur of the *same* species is thinner towards the south of the range of the same species than to the north—that the *same* shells are brighter-coloured to the south than north; that the same [shell] is paler-coloured in deep water—that insects are smaller and darker on mountains—more livid and testaceous near the sea—that plants are smaller and more hairy and with brighter flowers on mountains: now in all such, and other cases, distinct species in the two zones follow the same rule, which seems to me to be most simply explained by species, being only strongly marked varieties, and therefore following

* See 'Origin,' ed. i. p. 100.

the same laws as recognised and admitted varieties. I mention all this on account of the variation of plants in ascending mountains; I have quoted the foregoing remark only generally with no examples, for I add, there is so much doubt and dispute what to call varieties; but yet I have stumbled on so many casual remarks on *varieties* of plants on mountains being so characterised, that I presume there is some truth in it. What think you? Do you believe there is *any* tendency in *varieties*, as *generally* so called, of plants to become more hairy, and with proportionally larger and brighter-coloured flowers in ascending a mountain?

I have been interested in my "weed garden," of 3 × 2 feet square: I mark each seedling as it appears, and I am astonished at the number that come up, and still more at the number killed by slugs, &c. Already 59 have been so killed; I expected a good many, but I had fancied that this was a less potent check than it seems to be, and I attributed almost exclusively to mere choking, the destruction of the seedlings. Grass-seedlings seem to suffer much less than exogens. . . .

C. Darwin to J. D. Hooker.

Moor Park, Farnham, [April (?) 1857.]

MY DEAR HOOKER,—Your letter has been forwarded to me here, where I am undergoing hydropathy for a fortnight, having been here a week, and having already received an amount of good which is quite incredible to myself and quite unaccountable. I can walk and eat like a hearty Christian, and even my nights are good. I cannot in the least understand how hydropathy can act as it certainly does on me. It dulls one's brain splendidly; I have not thought about a single species of any kind since leaving home. Your note has taken me aback; I thought the hairiness, &c., of Alpine *species* was generally admitted; I am sure I have seen it alluded to a score of times. Falconer was haranguing on it the other day to me. Meyen or Gay, or some such fellow (whom you would despise), I remember, makes some remark on Chilian Cordillera plants. Wimmer has written a little book on the same lines, and on *varieties* being so characterized in the Alps. But after writing to you, I confess I was staggered by finding one man (Moquin-Tandon, I think) saying that Alpine flowers are strongly inclined to be white, and Linnæus saying that cold makes plants *apetalous*, even the same species! Are Arctic plants often apetalous? My general belief from my compiling work is quite to agree with what you say about the little direct influence of climate; and I have just alluded to the hairiness of Alpine plants as an *exception*. The odoriferousness would be a good case for me if I knew of *varieties* being more odoriferous in dry habitats.

I fear that I have looked at the hairiness of Alpine plants as so generally acknowledged that I have not marked passages, so as at all to see what kind of evidence authors advance. I must confess, the other day, when I asked Falconer, whether he knew of *individual* plants losing or acquiring hairiness when transported, he did not. But now *this second*, my memory flashes on me, and I am certain I have somewhere got marked a case of hairy plants from the Pyrenees losing hairs when cultivated at Montpellier. Shall you think me very impudent if I tell you that I have sometimes thought that (quite independently of the present case), you are a little too hard on bad observers; that a remark made by a bad observer *cannot* be right; an observer who deserves to be damned, you would utterly damn. I feel entire deference to any remark you make out of your own head; but when in opposition to some poor devil, I somehow involuntarily feel not quite so much, but yet much deference for your opinion. I do not know in the least whether there is any truth in this my criticism against you, but I have often thought I would tell you it.

I am really very much obliged for your letter, for, though I intended to put only one sentence and that vaguely, I should probably have put that much too strongly.

Ever, my dear Hooker, yours most truly,

C. DARWIN.

P.S.—This note, as you see, has not anything requiring an answer.

The distribution of fresh-water molluscs has been a horrid incubus to me, but I think I know my way now ; when first hatched they are very active, and I have had thirty or forty crawl on a dead duck's foot ; and they cannot be jerked off, and will live fifteen and even twenty-four hours out of water.

[The following letter refers to the expedition of the Austrian frigate *Novara ;* Lyell had asked my father for suggestions.]

C. Darwin to C. Lyell.

Down, Feb. 11th [1857].

MY DEAR LYELL,—I was glad to see in the newspapers about the Austrian Expedition.   I have nothing to add geologically to my notes in the Manual.*   I do not know whether the Expedition is tied down to call at only fixed spots.   But if there be any choice or power in the scientific men to influence the places—this would be most desirable.   It is my most deliberate conviction that nothing would aid more, Natural History, than careful collecting and investigating *all the productions* of the most isolated islands, especially of the southern hemisphere.   Except Tristan d'Acunha and Ker- guelen Land, they are very imperfectly known ; and even at Kerguelen Land, how much there is to make out about the lignite beds, and whether there are signs of old Glacial action. Every sea-shell and insect and plant is of value from such spots.   Some one in the Expedition especially ought to have

* The article "Geology" in the Admiralty 'Manual of Scientific Enquiry.'

Hooker's New Zealand Essay.   What grand work to explore Rodriguez, with its fossil birds, and little known productions of every kind.   Again the Seychelles, which, with the Cocos so near, must be a remnant of some older land.   The outer island of Juan Fernandez is little known.   The investigation of these little spots by a band of naturalists would be grand ; St. Paul's and Amsterdam would be glorious, botanically, and geologically.   Can you not recommend them to get my ' Journal ' and ' Volcanic Islands ' on account of the Galapagos. If they come from the north it will be a shame and a sin if they do not call at Cocos Islet, one of the Galapagos.   I always regretted that I was not able to examine the great craters on Albemarle Island, one of the Galapagos.   In New Zealand urge on them to look out for erratic boulders and marks of old glaciers.

Urge the use of the dredge in the Tropics ; how little or nothing we know of the limit of life downward in the hot seas ?

My present work leads me to perceive how much the domestic animals have been neglected in out of the way countries.

The Revillagigedo Island off Mexico, I believe, has never been trodden by foot of naturalist.

If the expedition sticks to such places as Rio, Cape of Good Hope, Ceylon and Australia, &c., it will not do much.

Ever yours most truly,

C. DARWIN.

[The following passage occurs in a letter to Mr. Fox, February 22, 1857, and has reference to the book on Evolution on which he was still at work :

"I am got most deeply interested in my subject ; though I wish I could set less value on the bauble fame, either present or posthumous, than I do, but not I think, to any extreme degree: yet, if I know myself, I would work just as hard, though with less gusto, if I knew that my book would be published for ever anonymously."]

C. Darwin to A. R. Wallace.

Moor Park, May 1st, 1857.

MY DEAR SIR,—I am much obliged for your letter of October 10th, from Celebes, received a few days ago; in a laborious undertaking, sympathy is a valuable and real encouragement. By your letter and even still more by your paper * in the Annals, a year or more ago, I can plainly see that we have thought much alike and to a certain extent have come to similar conclusions. In regard to the Paper in the Annals, I agree to the truth of almost every word of your paper; and I dare say that you will agree with me that it is very rare to find oneself agreeing pretty closely with any theoretical paper; for it is lamentable how each man draws his own different conclusions from the very same facts. This summer will make the 20th year (!) since I opened my first note-book, on the question how and in what way do species and varieties differ from each other. I am now preparing my work for publication, but I find the subject so very large, that though I have written many chapters, I do not suppose I shall go to press for two years. I have never heard how long you intend staying in the Malay Archipelago; I wish I might profit by the publication of your Travels there before my work appears, for no doubt you will reap a large harvest of facts. I have acted already in accordance with your advice of keeping domestic varieties, and those appearing in a state of nature, distinct; but I have sometimes doubted of the wisdom of this, and therefore I am glad to be backed by your opinion. I must confess, however, I rather doubt the truth

* "On the Law that has regulated the Introduction of New Species." —Ann. Nat. Hist., 1855.

of the now very prevalent doctrine of all our domestic animals having descended from several wild stocks ; though I do not doubt that it is so in some cases. I think there is rather better evidence on the sterility of hybrid animals than you seem to admit : and in regard to plants the collection of carefully recorded facts by Kölreuter and Gaertner (and Herbert) is *enormous*. I most entirely agree with you on the little effects of " climatal conditions," which one sees referred to *ad nauseam* in all books : I suppose some very little effect must be attributed to such influences, but I fully believe that they are very slight. It is really *impossible* to explain my views (in the compass of a letter), on the causes and means of variation in a state of nature ; but I have slowly adopted a distinct and tangible idea,—whether true or false others must judge ; for the firmest conviction of the truth of a doctrine by its author, seems, alas, not to be the slightest guarantee of truth ! . . .

C. *Darwin to J. D. Hooker.*

Moor Park, Saturday [May 2nd, 1857].

MY DEAR HOOKER,—You have shaved the hair off the Alpine plants pretty effectually. The case of the Anthyllis will make a " tie " with the believed case of Pyrenees plants becoming glabrous at low levels. If I *do* find that I have marked such facts, I will lay the evidence before you. I wonder how the belief could have originated ! Was it through final causes to keep the plants warm ? Falconer in talk coupled the two facts of woolly Alpine plants and mammals. How candidly and meekly you took my Jeremiad on your severity to second-class men. After I had sent it off, an ugly little voice asked me, once or twice, how much of my noble defence of the poor in spirit and in fact, was owing to your having not seldom smashed favourite notions of my own. I silenced the ugly little voice with contempt, but it would whisper again and again. I sometimes despise myself as a poor compiler as heartily as you could do, though I do *not* despise my whole work, as I think there is enough known to lay a foundation for the discussion on the origin of species. I have been led to despise and laugh at myself as a compiler, for having put down that "Alpine plants have large flowers," and now perhaps I may write over these very words, "Alpine plants have small or apetalous flowers!" . . .

C. Darwin to J. D. Hooker.

Down [May] 16th [1857].

MY DEAR HOOKER,—You said—I hope honestly—that you did not dislike my asking questions on general points, you of course answering or not as time and inclination might serve. I find in the animal kingdom that . . . . any part or organ developed normally, (*i.e.* not a monstrosity) in a species in any *high* or *unusual* degree, compared with the same part or organ in allied species, tends to be *highly variable*. I cannot doubt this from my mass of collected facts. To give an instance, the Cross-bill is very abnormal in the structure of its bill compared with other allied Fringillidæ, and the beak is *eminently variable*. The Himantopus, remarkable from the wonderful length of its legs, is *very* variable in the length of its legs. I could give *many* most striking and curious illustrations in all classes; so many that I think it cannot be chance. But I have *none* in the vegetable kingdom, owing, as I believe, to my ignorance. If Nepenthes consisted of *one* or two species in a group with a pitcher developed, then I should have expected it to have been very variable; but I do not consider Nepenthes a case in point, for when a whole genus or group has an organ, however anomalous, I do not expect it to be variable,— it is only when one or few species differ greatly in some one part or organ from the forms *closely allied* to it in all other respects, that I believe such part or organ to be highly vari-

able. Will you turn this in your mind? it is an important
apparent *law* (!) for me.

> Ever yours,
> C. DARWIN.

P.S.—I do not know how far you will care to hear, but
I find Moquin-Tandon treats in his 'Tératologie' on villosity
of plants, and seems to attribute more to dryness than
altitude ; but seems to think that it must be admitted that
mountain plants are villose, and that this villosity is only
in part explained by De Candolle's remark that the dwarfed
condition of mountain plants would condense the hairs, and
so give them the *appearance* of being more hairy. He quotes
Senebier, 'Physiologie Végétale,' as authority—I suppose
the first authority, for mountain plants being hairy.

If I could show positively that the endemic species were
more hairy in dry districts, then the case of the varieties
becoming more hairy in dry ground would be a fact for me.

C. Darwin to J. D. Hooker.

> Down, June 3rd [1857].

MY DEAR HOOKER,—I am going to enjoy myself by
having a prose on my own subjects to you, and this is a
greater enjoyment to me than you will readily understand, as
I for months together do not open my mouth on Natural
History. Your letter is of great value to me, and staggers me
in regard to my proposition. I dare say the absence of
botanical facts may in part be accounted for by the difficulty
of measuring slight variations. Indeed, after writing, this
occurred to me ; for I have *Crucianella stylosa* coming into
flower, and the pistil ought to be very variable in length, and
thinking of this I at once felt how could one judge whether it
was variable in any high degree. How different, for instance,
from the beak of a bird ! But I am not satisfied with this explanation, and am staggered. Yet I think there is something

in the law ; I have had so many instances, as the following :
I wrote to Wollaston to ask him to run through the Madeira
Beetles and tell me whether any one presented anything very
anomalous in relation to its allies.   He gave me a unique case
of an enormous head in a female, and then I found in his book,
already stated, that the size of the head was *astonishingly*
variable.   Part of the difference with plants may be accounted
for by many of my cases being secondary male or *female*
characters but then I have striking cases with hermaphrodite
Cirripedes.   The cases seem to me far too numerous for
accidental coincidences of great variability and abnormal development.   I presume that you will not object to my putting a note saying that you had reflected over the case, and
though one or two cases seemed to support, quite as many
or more seemed wholly contradictory.   This want of evidence is
the more surprising to me, as generally I find any proposition
more easily tested by observations in botanical works, which
I have picked up, than in zoological works.   I never dreamed
that you had kept the subject at all before your mind.   Altogether the case is one more of my *many* horrid puzzles.   My
observations, though on so infinitely a small scale, on the
struggle for existence, begin to make me see a little clearer
how the fight goes on.   Out of sixteen kinds of seed sown on
my meadow, fifteen have germinated, but now they are
perishing at such a rate that I doubt whether more than one
will flower.   Here we have choking which has taken place
likewise on a great scale, with plants not seedlings, in a bit of
my lawn allowed to grow up.   On the other hand, in a bit of
ground, 2 by 3 feet, I have daily marked each seedling weed
as it has appeared during March, April and May, and 357 have
come up, and of these 277 have *already* been killed, chiefly by
slugs.   By the way, at Moor Park, I saw rather a pretty case
of the effects of animals on vegetation : there are enormous
commons with clumps of old Scotch firs on the hills, and
about eight or ten years ago some of these commons were

enclosed, and all round the clumps nice young trees are springing up by the million, looking exactly as if planted, so many are of the same age. In other parts of the common, not yet enclosed, I looked for miles and not *one* young tree could be seen. I then went near (within quarter of a mile of the clumps) and looked closely in the heather, and there I found tens of thousands of young Scotch firs (thirty in one square yard) with their tops nibbled off by the few cattle which occasionally roam over these wretched heaths. One little tree, three inches high, by the rings appeared to be twenty-six years old, with a short stem about as thick as a stick of sealing-wax. What a wondrous problem it is, what a play of forces, determining the kind and proportion of each plant in a square yard of turf! It is to my mind truly wonderful. And yet we are pleased to wonder when some animal or plant becomes extinct.

I am so sorry that you will not be at the Club. I see Mrs. Hooker is going to Yarmouth ; I trust that the health of your children is not the motive. Good-bye.

My dear Hooker, ever yours,

C. DARWIN.

P.S.—I believe you are afraid to send me a ripe Edwardsia pod, for fear I should float it from New Zealand to Chile ! ! !

C. Darwin to J. D. Hooker.

Down, June 5 [1857].

MY DEAR HOOKER,—I honour your conscientious care about the medals.* Thank God! I am only an amateur (but a much interested one) on the subject.

It is an old notion of mine that more good is done by giving medals to younger men in the early part of their career, than as a mere reward to men whose scientific career is nearly finished. Whether medals ever do any good is a question which does

* The Royal Society's medals.

not concern us, as there the medals are.   I am almost inclined to think that I would rather lower the standard, and give medals to young workers than to old ones with no *especial* claims.   With regard to especial claims, I think it just deserving your attention, that if general claims are once admitted, it opens the door to great laxity in giving them. Think of the case of a very rich man, who aided *solely* with his money, but to a grand extent—or such an inconceivable prodigy as a minister of the Crown who really cared for science.   Would you give such men medals?   Perhaps medals could not be better applied than *exclusively* to such men.   I confess at present I incline to stick to especial claims which can be put down on paper. . . .

I am much confounded by your showing that there are not obvious instances of my (or rather Waterhouse's) law of abnormal developments being highly variable.   I have been thinking more of your remark about the difficulty of judging or comparing variability in plants from the great general variability of parts.   I should look at the law as more completely smashed if you would turn in your mind for a little while for cases of great variability of an organ, and tell me whether it is moderately easy to pick out such cases; *for if they can be picked out*, and, notwithstanding, do not coincide with great or abnormal development, it would be a complete smasher.   It is only beginning in your mind at the variability end of the question instead of at the abnormality end.   *Perhaps* cases in which a part is highly variable in all the species of a group should be excluded, as possibly being something distinct, and connected with the perplexing subject of polymorphism.   Will you perfect your assistance by further considering, for a little, the subject this way?

I have been so much interested this morning in comparing all my notes on the variation of the several species of the genus Equus and the results of their crossing.   Taking most strictly analogous facts amongst the blessed pigeons for my guide,

I believe I can plainly see the colouring and marks of the grandfather of the Ass, Horse, Quagga, Hemionus and Zebra, some millions of generations ago! Should not I [have] sneer[ed] at any one who made such a remark to me a few years ago; but my evidence seems to me so good that I shall publish my vision at the end of my little discussion on this genus.

I have of late inundated you with my notions, you best of friends and philosophers.

Adios,

C. DARWIN.

C. Darwin to J. D. Hooker.

Moor Park, Farnham, June 25th [1857].

MY DEAR HOOKER,—This requires no answer, but I will ask you whenever we meet. Look at enclosed seedling gorses, especially one with the top knocked off. The leaves succeeding the cotyledons being almost clover-like in shape, seems to me feebly analogous to embryonic resemblances in young animals, as, for instance, the young lion being striped. I shall ask you whether this is so.* . . .

Dr. Lane† and wife, and mother-in-law, Lady Drysdale, are some of the nicest people I have ever met.

I return home on the 30th. Good-bye, my dear Hooker.

Ever yours,

C. DARWIN.

[Here follows a group of letters, of various dates, bearing on the question of large genera varying.]

C. Darwin to J. D. Hooker.

March 11th [1858].

. . . I was led to all this work by a remark of Fries, that the species in large genera were more closely related to each

---

* See 'Power of Movements in Plants,' p. 414.
† The physician at Moor Park.

other than in small genera; and if this were so, seeing that varieties and species are so hardly distinguishable, I concluded that I should find more varieties in the large genera than in the small. . . . Some day I hope you will read my short discussion on the whole subject. You have done me infinite service, whatever opinion I come to, in drawing my attention to at least the possibility or the probability of botanists recording more varieties in the large than in the small genera. It will be hard work for me to be candid in coming to my conclusion.

Ever yours, most truly,

C. DARWIN.

P.S.—I shall be several weeks at my present job. The work has been turning out badly for me this morning, and I am sick at heart; and, oh! how I do hate species and varieties.

C. Darwin to J. D. Hooker.

July 14th [1857?]

. . . I write now to supplicate most earnestly a favour, viz. the loan of Boreau, *Flore du centre de la France, either 1st or 2nd edition*, last best; also "Flora Ratisbonensis," by Dr. Fürnrohr, in 'Naturhist. Topographie von Regensburg, 1839.' If you can *possibly* spare them, will you send them at once to the enclosed address. If you have not them, will you send one line by return of post: as I must try whether Kippist * can anyhow find them, which I fear will be nearly impossible in the Linnean Library, in which I know they are.

I have been making some calculations about varieties, &c., and talking yesterday with Lubbock, he has pointed out to me the grossest blunder which I have made in principle, and which entails two or three weeks' lost work; and I am at a dead-lock till I have these books to go over again, and see

---

* The late Mr. Kippist was at this time in charge of the Linnean Society's Library.

what the result of calculation on the right principle is. I am the most miserable, bemuddled, stupid dog in all England, and am ready to cry with vexation at my blindness and presumption.

Ever yours, most miserably,

C. DARWIN.

C. Darwin to John Lubbock.

Down, [July] 14th [1857].

MY DEAR LUBBOCK,—You have done me the greatest possible service in helping me to clarify my brains. If I am as muzzy on all subjects as I am on proportion and chance, —what a book I shall produce!

I have divided the New Zealand Flora as you suggested. There are 339 species in genera of 4 and upwards, and 323 in genera of 3 and less.

The 339 species have 51 species presenting one or more varieties. The 323 species have only 37. Proportionately (339 : 323 :: 51 : 48·5) they ought to have had 48½ species presenting vars. So that the case goes as I want it, but not strong enough, without it be general, for me to have much confidence in. I am quite convinced yours is the right way: I had thought of it, but should never have done it had it not been for my most fortunate conversation with you.

I am quite shocked to find how easily I am muddled, for I had before thought over the subject much, and concluded my way was fair. It is dreadfully erroneous.

What a disgraceful blunder you have saved me from. I heartily thank you.

Ever yours,

C. DARWIN.

P.S.—It is enough to make me tear up all my MS. and give up in despair.

It will take me several weeks to go over all my materials. But oh, if you knew how thankful I am to you!

C. Darwin to J. D. Hooker.

Down, Aug. [1857].

MY DEAR HOOKER,—It is a horrid bore you cannot come soon, and I reproach myself that I did not write sooner. How busy you must be! with such a heap of botanists at Kew. Only think, I have just had a letter from ·Henslow, saying he will come here between 11th and 15th! Is not that grand? Many thanks about Fürnrohr. I must humbly supplicate Kippist to search for it : he most kindly got Boreau for me.

I am got extremely interested in tabulating, according to mere size of genera, the species having any varieties marked by Greek letters or otherwise : the result (as far as I have yet gone) seems to me one of the most important arguments I have yet met with, that varieties are only small species—or species only strongly marked varieties. The subject is in many ways so very important for me ; I wish much you would think of any well-worked Floras with from 1000–2000 species, with the varieties marked. It is good to have hair-splitters and lumpers.*   I have done, or am doing :—

Babington	} British Flora.	
Henslow		
London Catalogue. H. C. Watson .		
Boreau	France.	
Miquel	Holland.	
Asa Gray	U. States.	
Hooker	{ N. Zealand.	
	{ Fragment of Indian Flora.	
Wollaston	Madeira insects.	

Has not Koch published a good German Flora? Does he mark varieties? Could you send it me? Is there not some grand Russian Flora, which perhaps has varieties marked? The Floras ought to be well known.

* Those who make many species are the " splitters," and those who make few are the " lumpers."

I am in no hurry for a few weeks. Will you turn this in your head, when, if ever, you have leisure? The subject is very important for my work, though I clearly see *many* causes of error. . . .

C. Darwin to Asa Gray.

Down, Feb. 21st [1859].

MY DEAR GRAY,—My last letter begged no favour, this one does: but it will really cost you very little trouble to answer me, and it will be of very *great* service to me, owing to a remark made to me by Hooker, which I cannot credit, and which was suggested to him by one of my letters. He suggested my asking you, and I told him I would not give the least hint what he thought. I generally believe Hooker implicitly, but he is sometimes, I think, and he confesses it, rather over-critical, and his ingenuity in discovering flaws seems to me admirable. Here is my question :—
" Do you think that good botanists in drawing up a local Flora, whether small or large, or in making a Prodromus like De Candolle's, would almost universally, but unintentionally and unconsciously, tend to record (*i.e.* marking with Greek letters and giving short characters) varieties in the large or in the small genera? Or would the tendency be to record the varieties about equally in genera of all sizes? Are you yourself conscious on reflection that you have attended to, and recorded more carefully the varieties in large or small, or very small genera ? "

I know what fleeting and trifling things varieties very often are; but my query applies to such as have been thought worth marking and recording. If you could screw time to send me ever so brief an answer to this, pretty soon, it would be a great service to me.

Yours most truly obliged,

CH. DARWIN.

P.S.—Do you know whether any one has ever published any remarks on the geographical range of varieties of plants in comparison with the species to which they are supposed to belong? I have in vain tried to get some vague idea, and with the exception of a little information on this head given me by Mr. Watson in a paper on Land Shells in U. States, I have quite failed ; but perhaps it would be difficult for you to give me even a brief answer on this head, and if so I am not so unreasonable, *I assure you*, as to expect it.

If you are writing to England soon, you could enclose other letters [for] me to forward.

Please observe, the question is not whether there are more or fewer varieties in larger or smaller genera, but whether there is a stronger or weaker tendency in the minds of botanists to *record* such in large or small genera.

C. Darwin to J. D. Hooker.

Down, May 6th [1858].

. . . I send by this post my MS. on the "commonness," "range," and "variation" of species in large and small genera. You have undertaken a horrid job in so very kindly offering to read it, and I thank you warmly. I have just corrected the copy, and am disappointed in finding how tough and obscure it is ; but I cannot make it clearer, and at present I loathe the very sight of it. The style of course requires further correction, and if published I must try, but as yet see not how, to make it clearer.

If you have much to say and can have patience to consider the whole subject, I would meet you in London on the Phil. Club day, so as to save you the trouble of writing. For Heaven's sake, you stern and awful judge and sceptic, remember that my conclusions may be true, notwithstanding that Botanists may have recorded more varieties in large than in small genera. It seems to me a mere balancing of probabilities. Again I thank you most sincerely, but I fear you will find it a horrid job.

Ever yours,
C. DARWIN.

[The letters now continue the history of the years 1857 and 1858.]

C. Darwin to A. R. Wallace.

Down, Dec. 22nd, 1857.

MY DEAR SIR,—I thank you for your letter of Sept. 27th. I am extremely glad to hear that you are attending to distribution in accordance with theoretical ideas. I am a firm believer that without speculation there is no good and original observation. Few travellers have attended to such points as you are now at work on ; and, indeed, the whole subject of distribution of animals is dreadfully behind that of plants. You say that you have been somewhat surprised at no notice having been taken of your paper in the Annals.* I cannot say that I am, for so very few naturalists care for anything beyond the mere description of species. But you must not suppose that your paper has not been attended to : two very good men, Sir C. Lyell, and Mr. E. Blyth at Calcutta, specially called my attention to it. Though agreeing with you on your conclusions in that paper, I believe I go much further than you ; but it is too long a subject to enter on my speculative notions. I have not yet seen your paper on the distribution of animals in the Aru Islands. I shall read it with the utmost interest ; for I think that the most interesting quarter of the whole globe in respect to distribution, and I have long been very imperfectly trying to collect data for the Malay Archipelago. I shall be quite prepared to subscribe to your

* "On the Law that has regulated the Introduction of New Species." —Ann. Nat. Hist., 1855.

doctrine of subsidence; indeed, from the quite independent evidence of the Coral Reefs I coloured my original map (in my Coral volume) of the Aru Islands as one of subsidence, but got frightened and left it uncoloured. But I can see that you are inclined to go much further than I am in regard to the former connection of oceanic islands with continents. Ever since poor E. Forbes propounded this doctrine, it has been eagerly followed; and Hooker elaborately discusses the former connection of all the Antarctic Islands and New Zealand and South America. About a year ago I discussed this subject much with Lyell and Hooker (for I shall have to treat of it), and wrote out my arguments in opposition; but you will be glad to hear that neither Lyell nor Hooker thought much of my arguments. Nevertheless, for once in my life, I dare withstand the almost preternatural sagacity of Lyell.

You ask about land-shells on islands far distant from continents: Madeira has a few identical with those of Europe, and here the evidence is really good, as some of them are sub-fossil. In the Pacific Islands there are cases of identity, which I cannot at present persuade myself to account for by introduction through man's agency; although Dr. Aug. Gould has conclusively shown that many land-shells have thus been distributed over the Pacific by man's agency. These cases of introduction are most plaguing. Have you not found it so in the Malay Archipelago? It has seemed to me in the lists of mammals of Timor and other islands, that *several* in all probability have been naturalised. . . .

You ask whether I shall discuss "man." I think I shall avoid the whole subject, as so surrounded with prejudices; though I fully admit that it is the highest and most interesting problem for the naturalist. My work, on which I have now been at work more or less for twenty years, will not fix or settle anything; but I hope it will aid by giving a large collection of facts, with one definite end. I get on very slowly, partly from ill-health, partly from being a very slow worker.

I have got about half written ; but I do not suppose I shall publish under a couple of years. I have now been three whole months on one chapter on Hybridism !

I am astonished to see that you expect to remain out three or four years more. What a wonderful deal you will have seen, and what interesting areas—the grand Malay Archipelago and the richest parts of South America ! I infinitely admire and honour your zeal and courage in the good cause of Natural Science ; and you have my very sincere and cordial good wishes for success of all kinds, and may all your theories succeed, except that on Oceanic Islands, on which subject I will do battle to the death.

Pray believe me, my dear sir, yours very sincerely,

C. DARWIN.

C. Darwin to W. D. Fox.

Feb. 8th [1858].

. . . I am working very hard at my book, perhaps too hard. It will be very big, and I am become most deeply interested in the way facts fall into groups. I am like Crœsus overwhelmed with my riches in facts, and I mean to make my book as perfect as ever I can. I shall not go to press at soonest for a couple of years. . . .

C. Darwin to J. D. Hooker.

Feb. 23rd [1858].

. . . I was not much struck with the great Buckle, and I admired the way you stuck up about deduction and induction. I am reading his book,* which, with much sophistry, as it seems to me, is *wonderfully* clever and original, and with astounding knowledge.

I saw that you admired Mrs. Farrer's 'Questa tomba' of

* 'The History of Civilisation.'

Beethoven thoroughly; there is something grand in her sweet tones.

Farewell. I have partly written this note to drive bee's-cells out of my head; for I am half-mad on the subject to try to make out some simple steps from which all the wondrous angles may result.*

I was very glad to see Mrs. Hooker on Friday; how well she appears to be and looks.

Forgive your intolerable but affectionate friend,

C. DARWIN.

C. Darwin to W. D. Fox.

Down, April 16th [1858].

MY DEAR FOX,—I want you to observe one point for me, on which I am extremely much interested, and which will give you no trouble beyond keeping your eyes open, and that is a habit I know full well that you have.

I find horses of various colours often have a spinal band or stripe of different and darker tint than the rest of the body; rarely transverse bars on the legs, generally on the under-side of the front legs, still more rarely a very faint transverse shoulder-stripe like an ass.

Is there any breed of Delamere forest ponies? I have found out little about ponies in these respects. Sir P. Egerton has, I believe, some quite thoroughbred chestnut horses; have any of them the spinal stripe? Mouse-coloured ponies, or rather small horses, often have spinal and leg bars. So have dun horses (by dun I mean real colour of cream mixed with brown, bay, or chestnut). So have sometimes chestnuts, but I have not yet got a case of spinal stripe in chestnut, race horse, or in quite heavy cart-horse. Any fact of this nature of such stripes in horses would be *most* useful to me. There is a

---

* He had much correspondence on this subject with the late Professor Miller of Cambridge.

parallel case in the legs of the donkey, and I have collected some most curious cases of stripes appearing in various crossed equine animals. I have also a large mass of parallel facts in the breeds of pigeons about the wing bars. I *suspect* it will throw light on the colour of the primeval horse. So do help me if occasion turns up. . . . My health has been lately very bad from overwork, and on Tuesday I go for a fortnight's hydropathy. My work is everlasting. Farewell.

My dear Fox, I trust you are well. Farewell,

C. DARWIN.

C. Darwin to J. D. Hooker.

Moor Park, Farnham [April 26th, 1858].

. . . I have just had the innermost cockles of my heart rejoiced by a letter from Lyell. I said to him (or he to me) that I believed from the character of the flora of the Azores, that icebergs must have been stranded there ; and that I expected erratic boulders would be detected embedded between the upheaved lava-beds ; and I got Lyell to write to Hartung to ask, and now H. says my question explains what had astounded him, viz. large boulders (and some polished) of mica-schist, quartz, sandstone, &c., some embedded, and some 40 and 50 feet above the level of the sea, so that he had inferred that they had not been brought as ballast. Is this not beautiful ?

The water-cure has done me some good, but I [am] nothing to boast of to-day, so good-bye.

My dear friend, yours,

C. D.

C. Darwin to C. Lyell.

Moor Park, Farnham, April 26th [1858].

MY DEAR LYELL,—I have come here for a fortnight's hydropathy, as my stomach had got, from steady work, into a

horrid state. I am extremely much obliged to you for sending me Hartung's interesting letter. The erratic boulders are
splendid. It is a grand case of floating ice versus glaciers.
He ought to have compared the northern and southern shores
of the islands. It is eminently interesting to me, for I have
written a very long chapter on the subject, collecting briefly
all the geological evidence of glacial action in different parts
of the world, and then at great length (on the theory of species
changing) I have discussed the migration and modification of
plants and animals, in sea and land, over a large part of the
world. To my mind, it throws a flood of light on the whole
subject of distribution, if combined with the modification of
species. Indeed, I venture to speak with some little confidence on this, for Hooker, about a year ago, kindly read
over my chapter, and though he then demurred gravely to
the general conclusion, I was delighted to hear a week or two
ago that he was inclined to come round pretty strongly to my
views of distribution and change during the glacial period. I
had a letter from Thompson, of Calcutta, the other day, which
helps me much, as he is making out for me what heat our
temperate plants can endure. But it is too long a subject for
a note ; and I have written thus only because Hartung's note
has set the whole subject afloat in my mind again. But I
will write no more, for my object here is to think about
nothing, bathe much, walk much, eat much, and read much
novels. Farewell, with many thanks, and very kind remembrance to Lady Lyell.

Ever yours,
C. DARWIN.

C. Darwin to Mrs. Darwin.

Moor Park, Wednesday, April [1858].

The weather is quite delicious. Yesterday, after writing to
you, I strolled a little beyond the glade for an hour and a half,

and enjoyed myself—the fresh yet dark-green of the grand
Scotch firs, the brown of the catkins of the old birches, with
their white stems, and a fringe of distant green from the
larches, made an excessively pretty view. At last I fell fast
asleep on the grass, and awoke with a chorus of birds singing
around me, and squirrels running up the trees, and some
woodpeckers laughing, and it was as pleasant and rural a
scene as ever I saw, and I did not care one penny how any of
the beasts or birds had been formed. I sat in the drawing-
room till after eight, and then went and read the Chief
Justice's summing up, and thought Bernard * guilty, and then
read a bit of my novel, which is feminine, virtuous, clerical,
philanthropical, and all that sort of thing, but very decidedly
flat. I say feminine, for the author is ignorant about money
matters, and not much of a lady—for she makes her men say,
"My Lady." I like Miss Craik very much, though we have
some battles, and differ on every subject. I like also the
Hungarian ; a thorough gentleman, formerly attaché at Paris,
and then in the Austrian cavalry, and now a pardoned exile,
with broken health. He does not seem to like Kossuth, but
says, he is certain [he is] a sincere patriot, most clever and
eloquent, but weak, with no determination of character. . . .

* Simon Bernard was tried in   Emperor of the French. The ver-
April 1858 as an accessory to   dict was "not guilty."
Orsini's attempt on the life of the

CHAPTER IV

THE WRITING OF THE 'ORIGIN OF SPECIES.

JUNE 18, 1858, TO NOVEMBER 1859.

[THE letters given in the present chapter tell their story with sufficient clearness, and need but a few words of explanation. Mr. Wallace's Essay, referred to in the first letter, bore the title, 'On the Tendency of Varieties to depart indefinitely from the Original Type,' and was published in the Linnean Society's 'Journal' (1858, vol. iii. p. 53) as part of the joint paper of " Messrs. C. Darwin and A. Wallace," of which the full title was 'On the Tendency of Species to form Varieties; and on the Perpetuation of Varieties and Species by Natural Means of Selection.'

My father's contribution of the paper consisted of (1) Extracts from the sketch of 1844; (2) part of a letter addressed to Dr. Asa Gray, dated September 5, 1857, and which is given at p. 120. The paper was "communicated" to the Society by Sir Charles Lyell and Sir Joseph Hooker, in whose prefatory letter, a clear account of the circumstances of the case is given.

Referring to Mr. Wallace's Essay, they wrote :—

" So highly did Mr. Darwin appreciate the value of the views therein set forth, that he proposed, in a letter to Sir Charles Lyell, to obtain Mr. Wallace's consent to allow the Essay to be published as soon as possible. Of this step we highly approved, provided Mr. Darwin did not withhold from the public, as he was strongly inclined to do (in favour of

Mr. Wallace), the memoir which he had himself written on the same subject, and which, as before stated, one of us had perused in 1844, and the contents of which we had both of us been privy to for many years.   On representing this to Mr. Darwin, he gave us permission to make what use we thought proper of his memoir, &c. ; and in adopting our present course, of presenting it to the Linnean Society, we have explained to him that we are not solely considering the relative claims to priority of himself and his friend, but the interests of science generally."]

LETTERS.

C. Darwin to C. Lyell.

Down, 18th [June 1858].

MY DEAR LYELL,—Some year or so ago you recommended me to read a paper by Wallace in the 'Annals,' * which had interested you, and, as I was writing to him, I knew this would please him much, so I told him.   He has to-day sent me the enclosed, and asked me to forward it to you.   It seems to me well worth reading.   Your words have come true with a vengeance—that I should be forestalled.   You said this, when I explained to you here very briefly my views of 'Natural Selection' depending on the struggle for existence.   I never saw a more striking coincidence ; if Wallace had my MS. sketch written out in 1842, he could not have made a better short abstract!   Even his terms now stand as heads of my chapters.   Please return me the MS., which he does not say he wishes me to publish, but I shall, of course, at once write and offer to send to any journal.   So all my originality, what-ever it may amount to, will be smashed, though my book,

* Annals and Mag. of Nat. Hist., 1855.

if it will ever have any value, will not be deteriorated ; as all the labour consists in the application of the theory.

I hope you will approve of Wallace's sketch, that I may tell him what you say.

My dear Lyell, yours most truly,

C. DARWIN.

C. Darwin to C. Lyell.

Down, Friday [June 25, 1858].

MY DEAR LYELL,—I am very sorry to trouble you, busy as you are, in so merely personal an affair ; but if you will give me your deliberate opinion, you will do me as great a service as ever man did, for I have entire confidence in your judgment and honour. . . .

There is nothing in Wallace's sketch which is not written out much fuller in my sketch, copied out in 1844, and read by Hooker some dozen years ago. About a year ago I sent a short sketch, of which I have a copy, of my views (owing to correspondence on several points) to Asa Gray, so that I could most truly say and prove that I take nothing from Wallace. I should be extremely glad now to publish a sketch of my general views in about a dozen pages or so ; but I cannot persuade myself that I can do so honourably. Wallace says nothing about publication, and I enclose his letter. But as I had not intended to publish any sketch, can I do so honourably, because Wallace has sent me an outline of his doctrine ? I would far rather burn my whole book, than that he or any other man should think that I had behaved in a paltry spirit. Do you not think his having sent me this sketch ties my hands ? . . . . If I could honourably publish, I would state that I was induced now to publish a sketch (and I should be very glad to be permitted to say, to follow your advice long ago given) from Wallace having sent me an outline of my general conclusions. We differ only, [in] that I was led to my views from what artificial selection has done for domestic animals. I would send Wallace a copy of my letter to Asa Gray, to show him that I had not stolen his doctrine. But I cannot tell whether to publish now would not be base and paltry. This was my first impression, and I should have certainly acted on it had it not been for your letter.

This is a trumpery affair to trouble you with, but you cannot tell how much obliged I should be for your advice.

By the way, would you object to send this and your answer to Hooker to be forwarded to me, for then I shall have the opinion of my two best and kindest friends. This letter is miserably written, and I write it now, that I may for a time banish the whole subject; and I am worn out with musing . . .

My good dear friend, forgive me. This is a trumpery letter, influenced by trumpery feelings.

Yours most truly,
C. DARWIN.

I will never trouble you or Hooker on the subject again.

C. Darwin to C. Lyell.

Down, 26th [June 1858].

MY DEAR LYELL,—Forgive me for adding a P.S. to make the case as strong as possible against myself.

Wallace might say, "You did not intend publishing an abstract of your views till you received my communication. Is it fair to take advantage of my having freely, though unasked, communicated to you my ideas, and thus prevent me forestalling you?" The advantage which I should take being that I am induced to publish from privately knowing that Wallace is in the field. It seems hard on me that I should be thus compelled to lose my priority of many years' standing, but I cannot feel at all sure that this alters the

justice of the case.   First impressions are generally right, and
I at first thought it would be dishonourable in me now to
publish.

Yours most truly,
C. DARWIN.

P.S.—I have always thought you would make a first-rate
Lord Chancellor; and I now appeal to you as a Lord
Chancellor.

C. Darwin to J. D. Hooker.

Down, Tuesday [June 29, 1858].

. . . . I have received your letters.   I cannot think now *
on the subject, but soon will.   But I can see that you have
acted with more kindness, and so has Lyell, even than I could
have expected from you both, most kind as you are.

I can easily get my letter to Asa Gray copied, but it is too
short.

. . . . God bless you.   You shall hear soon, as soon as I
can think.

Yours affectionately,
C. DARWIN.

C. Darwin to J. D. Hooker.

Tuesday night [June 29, 1858].

MY DEAR HOOKER,—I have just read your letter, and see
you want the papers at once.   I am quite prostrated, and
can do nothing, but I send Wallace, and the abstract † of my
letter to Asa Gray, which gives most imperfectly only the
means of change, and does not touch on reasons for believing
that species do change.   I dare say all is too late.   I hardly

---

* So soon after the death, from
scarlet fever, of his infant child.

† "Abstract" is here used in
the sense of "extract;" in this
sense also it occurs in the 'Linnean
Journal,' where the sources o my
father's paper are described.

care about it.   But you are too generous to sacrifice so much
time and kindness.   It is most generous, most kind.   I send
my sketch of 1844 solely that you may see by your own
handwriting that you did read it.   I really cannot bear to
look at it.   Do not waste much time.   It is miserable in me
to care at all about priority.

The table of contents will show what it is.

I would make a similar, but shorter and more accurate
sketch for the 'Linnean Journal.'

I will do anything.   God bless you, my dear kind friend.
I can write no more.   I send this by my servant to Kew.

Yours,

C. DARWIN.

[The following letter is that already referred to as forming
part of the joint paper published in the Linnean Society's
Journal,' 1858] :—

C. Darwin to Asa Gray.

Down, Sept.* 5th [1857].

MY DEAR GRAY,—I forget the exact words which I used
in my former letter, but I dare say I said that I thought you
would utterly despise me when I told you what views I had
arrived at, which I did because I thought I was bound as an
honest man to do so.   I should have been a strange mortal,
seeing how much I owe to your quite extraordinary kindness, if
in saying this I had meant to attribute the least bad feeling to
you.   Permit me to tell you that, before I had ever cor-
responded with you, Hooker had shown me several of your
letters (not of a private nature), and these gave me the
warmest feeling of respect to you; and I should indeed be

---

* The date is given as October
in the 'Linnean Journal.'   The
extracts were printed from a duplicate undated copy in my father's
possession, on which he had written,
" This was sent to Asa Gray 8 or 9
months ago, I think October 1857."

ungrateful if your letters to me, and all I have heard of you, had not strongly enhanced this feeling. But I did not feel in the least sure that when you knew whither I was tending, you might not think me so wild and foolish in my views (God knows, arrived at slowly enough, and I hope conscientiously), that you would think me worth no more notice or assistance. To give one example: the last time I saw my dear old friend Falconer, he attacked me most vigorously, but quite kindly, and told me, "You will do more harm than any ten Naturalists will do good. I can see that you have already *corrupted* and half-spoiled Hooker!!" Now when I see such strong feeling in my oldest friends, you need not wonder that I always expect my views to be received with contempt. But enough and too much of this.

I thank you most truly for the kind spirit of your last letter. I agree to every word in it, and think I go as far as almost any one in seeing the grave difficulties against my doctrine. With respect to the extent to which I go, all the arguments in favour of my notions fall *rapidly* away, the greater the scope of forms considered. But in animals, embryology leads me to an enormous and frightful range. The facts which kept me longest scientifically orthodox are those of adaptation—the pollen-masses in asclepias — the mistletoe, with its pollen carried by insects, and seed by birds—the woodpecker, with its feet and tail, beak and tongue, to climb the tree and secure insects. To talk of climate or Lamarckian habit producing such adaptations to other organic beings is futile. This difficulty I believe I have surmounted. As you seem interested in the subject, and as it is an *immense* advantage to me to write to you and to hear, ever so briefly, what you think, I will enclose (copied, so as to save you trouble in reading) the briefest abstract of my notions on the means by which Nature makes her species. Why I think that species have really changed, depends on general facts in the affinities, embryology, rudimentary organs, geological history, and geo-

graphical distribution of organic beings. In regard to my
Abstract, you must take immensely on trust, each paragraph
occupying one or two chapters in my book. You will,
perhaps, think it paltry in me, when I ask you not to mention
my doctrine ; the reason is, if any one, like the author of the
' Vestiges,' were to hear of them, he might easily work them
in, and then I should have to quote from a work perhaps
despised by naturalists, and this would greatly injure any
chance of my views being received by those alone whose
opinions I value. [Here follows a discussion on "large
genera varying," which has no direct connection with the
remainder of the letter.]

I. It is wonderful what the principle of Selection by Man,
that is the picking out of individuals with any desired quality,
and breeding from them, and again picking out, can do.
Even breeders have been astonished at their own, results.
They can act on differences inappreciable to an uneducated
eye. Selection has been *methodically* followed in Europe for
only the last half century. But it has occasionally, and even
in some degree methodically, been followed in the most
ancient times. There must have been also a kind of unconscious selection from the most ancient times, namely, in the
preservation of the individual animals (without any thought of
their offspring) most useful to each race of man in his particular circumstances. The "roguing," as nursery-men call the
destroying of varieties, which depart from their type, is a kind
of selection. I am convinced that intentional and occasional
selection has been the main agent in making our domestic
races. But, however this may be, its great power of modification has been indisputably shown in late times. Selection
acts only by the accumulation of very slight or greater
variations, caused by external conditions, or by the mere
fact that in generation the child is not absolutely similar to
its parent. Man, by this power of accumulating variations,
adapts living beings to his wants—he *may be said* to make.

the wool of one sheep good for carpets, and another for cloth, &c.

II. Now, suppose there was a being, who did not judge by mere external appearance, but could study the whole internal organisation—who never was capricious—who should go on selecting for one end during millions of generations, who will say what he might not effect! In nature we have some *slight* variations, occasionally in all parts: and I think it can be shown that a change in the conditions of existence is the main cause of the child not exactly resembling its parents; and in nature, geology shows us what changes have taken place, and are taking place. We have almost unlimited time: no one but a practical geologist can fully appreciate this: think of the Glacial period, during the whole of which the same species of shells at least have existed; there must have been during this period, millions on millions of generations.

III. I think it can be shown that there is such an unerring power at work, or *Natural Selection* (the title of my book), which selects exclusively for the good of each organic being. The elder De Candolle, W. Herbert, and Lyell, have written strongly on the struggle for life; but even they have not written strongly enough. Reflect that every being (even the elephant) breeds at such a rate that, in a few years, or at most a few centuries or thousands of years, the surface of the earth would not hold the progeny of any one species. I have found it hard constantly to bear in mind that the increase of every single species is checked during some part of its life, or during some shortly recurrent generation. Only a few of those annually born can live to propagate their kind. What a trifling difference must often determine which shall survive and which perish!

IV. Now take the case of a country undergoing some change; this will tend to cause some of its inhabitants to vary slightly; not but what I believe most beings vary at all times

enough for selection to act on.   Some of its inhabitants will
be exterminated, and the remainder will be exposed to the
mutual action of a different set of inhabitants, which I believe
to be more important to the life of each being than mere
climate.   Considering the infinitely various ways beings have
to obtain food by struggling with other beings, to escape
danger at various times of life, to have their eggs or seeds
disseminated, &c. &c., I cannot doubt that during millions of
generations individuals of a species will be born with some
slight variation profitable to some part of its economy; such
will have a better chance of surviving, propagating this variation, which again will be slowly increased by the accumulative
action of natural selection ; and the variety thus formed will
either coexist with, or more commonly will exterminate its
parent form.   An organic being like the woodpecker, or
the mistletoe, may thus come to be adapted to a score of
contingencies ;   natural selection, accumulating those slight
variations in all parts of its structure which are in any way
useful to it, during any part of its life.

V.   Multiform difficulties will occur to every one on this
theory.   Most can, I think, be satisfactorily answered.—
" Natura non facit saltum " answer some of the most obvious.   The slowness of the change, and only a very few undergoing change at any one time answers others.   The extreme
imperfections of our geological records answer others.

VI.   One other principle, which may be called the principle
of divergence, plays, I believe, an important part in the origin
of species.   The same spot will support more life if occupied
by very diverse forms : we see this in the many generic forms
in a square yard of turf (I have counted twenty species
belonging to eighteen genera), or in the plants and insects,
on any little uniform islet, belonging to almost as many
genera and families as to species.   We can understand this
with the higher animals, whose habits we best understand.
We know that it has been experimentally shown that a plot

of land will yield a greater weight, if cropped with several species of grasses, than with two or three species. Now every single organic being, by propagating rapidly, may be said to be striving its utmost to increase in numbers. So it will be with the offspring of any species after it has broken into varieties, or sub-species, or true species. And it follows, I think, from the foregoing facts, that the varying offspring of each species will try (only few will succeed) to seize on as many and as diverse places in the economy of nature as possible. Each new variety or species when formed will generally take the place of, and so exterminate its less well-fitted parent. This, I believe, to be the origin of the classification or arrangement of all organic beings at all times. These always *seem* to branch and sub-branch like a tree from a common trunk; the flourishing twigs destroying the less vigorous—the dead and lost branches rudely representing extinct genera and families.

This sketch is *most* imperfect; but in so short a space I cannot make it better. Your imagination must fill up many wide blanks. Without some reflection, it will appear all rubbish; perhaps it will appear so after reflection.

C. D.

P.S.—This little abstract touches only the accumulative power of natural selection, which I look at as by far the most important element in the production of new forms. The laws governing the incipient or primordial variation (unimportant except as the groundwork for selection to act on, in which respect it is all important), I shall discuss under several heads, but I can come, as you may well believe, only to very partial and imperfect conclusions.

[The joint paper of Mr. Wallace and my father was read at the Linnean Society on the evening of July 1st. Sir Charles Lyell and Sir J. D. Hooker were present, and both, I believe, made a few remarks, chiefly with a view of impressing on those present the necessity of giving the most careful consideration to what they had heard.   There was, however, no semblance of a discussion.   Sir Joseph Hooker writes to me : " The interest excited was intense, but the subject was too novel and too ominous for the old school to enter the lists, before armouring.   After the meeting it was talked over with bated breath : Lyell's approval, and perhaps in a small way mine, as his lieutenant in the affair, rather overawed the Fellows, who would otherwise have flown out against the doctrine. We had, too, the vantage ground of being familiar with the authors and their theme."]

C. Darwin to J. D. Hooker.

Down, July 5th [1858].

MY DEAR HOOKER,—We are become more happy and less panic-struck, now that we have sent out of the house every child, and shall remove H., as soon as she can move. The first nurse became ill with ulcerated throat and quinsy, and the second is now ill with the scarlet fever, but, thank God, is recovering.   You may imagine how frightened we have been.   It has been a most miserable fortnight.   Thank you much for your note, telling me that all had gone on prosperously at the Linnean Society.   You must let me once again tell you how deeply I feel your generous kindness and Lyell's on this occasion.   But in truth it shames me that you should have lost time on a mere point of priority.   I shall be curious to see the proofs.   I do not in the least understand whether my letter to A. Gray is to be printed ; I suppose not, only your note ; but I am quite indifferent, and place myself absolutely in your and Lyell's hands.

I can easily prepare an abstract of my whole work, but I can hardly see how it can be made scientific for a Journal, without giving facts, which would be impossible.   Indeed, a mere abstract cannot be very short.   Could you give me any idea how many pages of the Journal could probably be spared me?

Directly after my return home, I would begin and cut my cloth to my measure. If the Referees were to reject it as not strictly scientific, I could, perhaps, publish it as a pamphlet.

With respect to my big interleaved abstract,* would you send it any time before you leave England, to the enclosed address? If you do not go till August 7th–10th, I should prefer it left with you. I hope you have jotted criticisms on my MS. on big Genera, &c., sufficient to make you remember your remarks, as I should be infinitely sorry to lose them. And I see no chance of our meeting if you go soon abroad. We thank you heartily for your invitation to join you: I can fancy nothing which I should enjoy more; but our children are too delicate for us to leave; I should be mere living lumber.

Lastly, you said you would write to Wallace; I certainly should much like this, as it would quite exonerate me: if you would send me your note, sealed up, I would forward it with my own, as I know the address, &c.

Will you answer me some time about your notions of the length of my abstract.

If you see Lyell, will you tell him how truly grateful I feel for his kind interest in this affair of mine. You must know that I look at it, as very important, for the reception of the view of species not being immutable, the fact of the greatest Geologist and Botanist in England taking *any sort of interest* in the subject: I am sure it will do much to break down prejudices.

Yours affectionately,

C. DARWIN.

* The Sketch of 1844.

C. Darwin to J. D. Hooker.

Miss Wedgwood's, Hartfield, Tunbridge Wells,
[July 13th, 1858].

MY DEAR HOOKER,—Your letter to Wallace seems to me perfect, quite clear and most courteous. I do not think it could possibly be improved, and I have to-day forwarded it with a letter of my own. I always thought it very possible that I might be forestalled, but I fancied that I had a grand enough soul not to care ; but I found myself mistaken and punished ; I had, however, quite resigned myself, and had written half a letter to Wallace to give up all priority to him, and should certainly not have changed had it not been for Lyell's and your quite extraordinary kindness. I assure you I feel it, and shall not forget it. I am *more* than satisfied at what took place at the Linnean Society. I had thought that your letter and mine to Asa Gray were to be only an appendix to Wallace's paper.

We go from here in a few days to the sea-side, probably to the Isle of Wight, and on my return (after a battle with pigeon skeletons) I will set to work at the abstract, though how on earth I shall make anything of an abstract in thirty pages of the Journal, I know not, but will try my best. I shall order Bentham ; is it not a pity that you should waste time in tabulating varieties? for I can get the Down schoolmaster to do it on my return, and can tell you all the results.

I must try and see you before your journey ; but do not think I am fishing to ask you to come to Down, for you will have no time for that.

You cannot imagine how pleased I am that the notion of Natural Selection has acted as a purgative on your bowels of immutability. Whenever naturalists can look at species changing as certain, what a magnificent field will be open,— on all the laws of variation,—on the genealogy of all living beings,—on their lines of migration, &c. &c. Pray thank

Mrs. Hooker for her very kind little note, and pray say how truly obliged I am, and in truth ashamed to think that she should have had the trouble of copying my ugly MS. It was extraordinarily kind in her. Farewell, my dear kind friend.

Yours affectionately,

C. DARWIN.

P.S.—I have had some fun here in watching a slave-making ant; for I could not help rather doubting the wonderful stories, but I have now seen a defeated marauding party, and I have seen a migration from one nest to another of the slave-makers, carrying their slaves (who are *house*, and not field niggers) in their mouths!

I am inclined to think that it is a true generalisation that, when honey is secreted at one point of the circle of the corolla, if the pistil bends, it always bends into the line of the gangway to the honey. The Larkspur is a good instance, in contrast to Columbine,—if you think of it, just attend to this little point.

C. Darwin to C. Lyell.

King's Head Hotel, Sandown, Isle of Wight.
July 18th [1858].

. . . We are established here for ten days, and then go on to Shanklin, which seems more amusing to one, like myself, who cannot walk. We hope much that the sea may do H. and L. good. And if it does, our expedition will answer, but not otherwise.

I have never half thanked you for all the extraordinary trouble and kindness you showed me about Wallace's affair. Hooker told me what was done at the Linnean Society, and I am far more than satisfied, and I do not think that Wallace can think my conduct unfair in allowing you and Hooker to do whatever you thought fair. I certainly was a little annoyed to lose all priority, but had resigned myself to my fate. I am going to prepare a longer abstract; but it is really impossible to do justice to the subject, except by giving the facts on which each conclusion is grounded, and that will, of course, be absolutely impossible. Your name and Hooker's name appearing as in any way the least interested in my work will, I am certain, have the most important bearing in leading people to consider the subject without prejudice. I look at this as so very important, that I am almost glad of Wallace's paper for having led to this.

My dear Lyell, yours most gratefully,

CH. DARWIN.

[The following letter refers to the proof-sheets of the Linnean paper. The 'introduction' means the prefatory letter signed by Sir C. Lyell and Sir J. D. Hooker.]

C. Darwin to J. D. Hooker.

King's Head Hotel, Sandown, Isle of Wight.

July 21st [1858].

MY DEAR HOOKER,—I received only yesterday the proof-sheets, which I now return. I think your introduction cannot be improved.

I am disgusted with my bad writing. I could not improve it, without rewriting all, which would not be fair or worth while, as I have begun on a better abstract for the Linnean Society. My excuse is that it *never* was intended for publication. I have made only a few corrections in the style; but I cannot make it decent, but I hope moderately intelligible. I suppose some one will correct the revise. (Shall I?)

Could I have a clean proof to send to Wallace?

I have not yet fully considered your remarks on big genera (but your general concurrence is of the *highest possible* interest to me); nor shall I be able till I re-read my MS.; but you may rely on it that you never make a remark to me which is lost from *inattention*. I am particularly glad you do not object to my stating your objections in a modified form, for they always struck me as very important, and as having much inherent value, whether or no they were fatal to my notions. I will consider and reconsider all your remarks. . . .

I have ordered Bentham, for, as —— says, it will be very curious to see a Flora written by a man who knows nothing of British plants!!

I am very glad at what you say about my Abstract, but you may rely on it that I will condense to the utmost. I would aid in money if it is too long.* In how many ways you have aided me!

Yours affectionately,

C. DARWIN.

[The 'Abstract' mentioned in the last sentence of the preceding letter was in fact the 'Origin of Species,' on which he now set to work. In his 'Autobiography' (p. 85) he speaks of beginning to write in September, but in his Diary he wrote, "July 20 to Aug. 12, at Sandown, began Abstract of Species book." "Sep. 16, Recommenced Abstract." The book was begun with the idea that it would be published as a paper, or series of papers, by the Linnean Society, and it was only in the late autumn that it became clear that it must take the form of an independent volume.]

C. Darwin to J. D. Hooker.

Norfolk House, Shanklin, Isle of Wight.
Friday [July] 30th [1858].

MY DEAR HOOKER,—Will you give the enclosed scrap to Sir William to thank him for his kindness ; and this gives me an excuse to amuse myself by writing to you a note, which requires no answer.

---

* That is to say, he would help to pay for the printing, if it should prove too long for the Linnean Society.

This is a very charming place, and we have got a very comfortable house. But, alas, I cannot say that the sea has done H. or L. much good. Nor has my stomach recovered from all our troubles. I am very glad we left home, for six children have now died of scarlet fever in Down. We return on the 14th of August.

I have got Bentham,* and am charmed with it, and William (who has just started for a tour abroad) has been making out all sorts of new (to me) plants capitally. The little scraps of information are so capital . . . The English names in the analytical keys drive us mad : give them by all means, but why on earth [not] make them subordinate to the Latin ; it puts me in a passion. W. charged into the Compositæ and Umbelliferæ like a hero, and demolished ever so many in grand style.

I pass my time by doing daily a couple of hours of my Abstract, and I find it amusing and improving work. I am now most heartily obliged to you and Lyell for having set me on this ; for I shall, when it is done, be able to finish my work with greater ease and leisure. I confess I hated the thought of the job ; and now I find it very unsatisfactory in not being able to give my reasons for each conclusion.

It will be longer than I expected ; it will take thirty-five of my MS. folio pages to give an abstract on variation under domestication alone ; but I will try to put in nothing which does not seem to me of some interest, and which was once new to me. It seems a queer plan to give an abstract of an unpublished work ; nevertheless, I repeat, I am extremely glad I have begun in earnest on it.

I hope you and Mrs. Hooker will have a very very pleasant tour. Farewell, my dear Hooker.

Yours affectionately,

C. DARWIN.

* 'British Flora.'

C. Darwin to J. D. Hooker.

Norfolk House, Shanklin, Isle of Wight.
Thursday [Aug. 5, 1858].

MY DEAR HOOKER,—I should think the note apologetical about the style of the Abstract was best as a note . . . . But I write now to ask you to send me by return of post the MS. on big genera, that I may make an abstract of a couple of pages in length. I presume that you have quite done with it, otherwise I would not for anything have it back. If you tie it with string, and mark it MS. for printing, it will not cost, I should think, more than 4*d.* I shall wish much to say that you have read this MS. and concur ; but you shall, before I read it to the Society, hear the sentence.

What you tell me after speaking with Busk about the length of the Abstract is an *immense* relief to me ; it will make the labour far less, not having to shorten so much every single subject ; but I will try not to be too diffusive. I fear it will spoil all interest in my book,* whenever published. The Abstract will do very well to divide into several parts : thus I have just finished "Variation under Domestication," in forty-four MS. pages, and that would do for one evening ; but I should be extremely sorry if all could not be published together.

What else you say about my Abstract pleases me highly, but frightens me, for I fear I shall never be able to make it good enough. But how I do run on about my own affairs to you !

I was astonished to see Sir W. Hooker's card here two or three days ago : I was unfortunately out walking. Henslow, also, has written to me, proposing to come to Down on the 9th, but alas, I do not return till the 13th, and my wife not till a week later ; so that I am also most sorry to think I shall

---

* The larger book begun in 1856.

not see you, for I should not like to leave home so soon.
I had thought of going to London and running down for an
hour or two to Kew. . . .

C. Darwin to J. D. Hooker.

Norfolk House, Shanklin, Isle of Wight.
[August 1858.]

MY DEAR HOOKER,—I write merely to say that the MS.
came safely two or three days ago.   I am much obliged for
the correction of style : I find it unutterably difficult to write
clearly.   When we meet I must talk over a few points on the
subject.

You speak of going to the sea-side somewhere ; we think
this the nicest sea-side place which we have ever seen,
and we like Shanklin better than other spots on the south
coast of the island, though many are charming and prettier,
so that I would suggest your thinking of this place.   We are
on the actual coast ; but tastes differ so much about places.

If you go to Broadstairs, when there is a strong wind from
the coast of France and in fine, dry, warm weather, look out
and you will *probably* (!) see thistle-seeds blown across the
Channel.   The other day I saw one blown right inland, and
then in a few minutes a second one and then a third ; and I
said to myself, God bless me, how many thistles there must be
in France ; and I wrote a letter in imagination to you.   But
I then looked at the *low* clouds, and noticed that they were
not coming inland, so I feared a screw was loose, I then walked
beyond a headland and found the wind parallel to the coast,
and on this very headland a noble bed of thistles, which by
every wide eddy were blown far out to sea, and then came
right in at right angles to the shore !   One day such a number
of insects were washed up by the tide, and I brought to life
thirteen species of Coleoptera ; not that I suppose these came
from France.   But do you watch for thistle-seed as you saunter
along the coast. . . .

C. Darwin to Asa Gray.

Aug. 11th [1858].

MY DEAR GRAY,—Your note of July 27th has just reached me in the Isle of Wight. It is a real and great pleasure to me to write to you about my notions ; and even if it were not so, I should be a most ungrateful dog, after all the invaluable assistance which you have rendered me, if I did not do anything which you asked.

I have discussed in my long MS. the later changes of climate and the effect on migration, and I will here give you an *abstract* of an *abstract* (which latter I am preparing of my whole work for the Linnean Society). I cannot give you facts, and I must write dogmatically, though I do not feel so on any point. I may just mention, in order that you may believe that I have *some* foundation for my views, that Hooker has read my MS., and though he at first demurred to my main point, he has since told me that further reflection and new facts have made him a convert.

In the older, or perhaps newer, Pliocene age (a little *before* the Glacial epoch) the temperature was higher ; of this there can be little doubt ; the land, on a *large scale*, held much its present disposition : the species were mainly, judging from shells, what they are now. At this period when all animals and plants ranged 10° or 15° nearer the poles, I believe the northern part of Siberia and of North America, being almost *continuous*, were peopled (it is quite possible, considering the shallow water, that Behring Straits were united, perhaps a little southward) by a nearly uniform fauna and flora, just as the Arctic regions now are. The climate then became gradually colder till it became what it now is ; and then the temperate parts of Europe and America would be separated, as far as migration is concerned, just as they now are. Then came on the Glacial period, driving far south all living things ; middle or even southern

Europe being peopled with Arctic productions; as the warmth
returned, the Arctic productions slowly crawled up the mountains as they became denuded of snow; and we now see on
their summits the remnants of a once continuous flora and
fauna.   This is E. Forbes's theory, which, however, I may
add, I had written out four years before he published.

Some facts have made me vaguely *suspect* that between
the glacial and the present temperature there was a period
of *slightly* greater warmth.   According to my modification-
doctrines, I look at many of the species of North America
which *closely* represent those of Europe, as having become
modified since the Pliocene period, when in the northern part
of the world there was nearly free communication between
the old and new worlds.   But now comes a more important
consideration; there is a considerable body of geological
evidence that during the Glacial epoch the whole world was
colder; I inferred that, many years ago, from erratic boulder
phenomena carefully observed by me on both the east and
west coast of South America.   Now I am so bold as to
believe that at the height of the Glacial epoch, *and when all
Tropical productions must have been considerably distressed,*
several temperate forms slowly travelled into the heart of the
Tropics, and even reached the southern hemisphere; and some
few southern forms penetrated in a reverse direction north-
ward.   (Heights of Borneo with Australian forms, Abyssinia
with Cape forms.) Wherever there was nearly continuous *high*
land, this migration would have been immensely facilitated;
hence the European character of the plants of Tierra del Fuego
and summits of Cordilleras; hence ditto on Himalaya. As the
temperature rose, all the temperate intruders would crawl up
the mountains.   Hence the European forms on Nilgherries,
Ceylon, summit of Java, Organ Mountains of Brazil. But
these intruders being surrounded with new forms would be
very liable to be improved or modified by natural selection,
to adapt them to the new forms with which they had to

compete; hence most of the forms on the mountains of the Tropics are not identical, but *representative* forms cf North temperate plants.

There are similar classes of facts in marine productions. All this will appear very rash to you, and rash it may be; but I am sure not so rash as it will at first appear to you: Hooker could not stomach it at all at first, but has become largely a,convert. From mammalia and shallow sea, I believe Japan to have been joined to main land of China within no remote period; and then the migration north and south before, during, and after the Glacial epoch would act on Japan, as on the corresponding latitude of China and the United States.

I should beyond anything like to know whether you have any Alpine collections from Japan, and what is their character. This letter is miserably expressed, but perhaps it will suffice to show what I believe have been the later main migrations and changes of temperature. . . .

C. Darwin to J. D. Hooker.

[Down,] Oct. 6th, 1858.

. . . If you have or can make leisure, I should very much like to hear news of Mrs. Hooker, yourself, and the children. Where did you go, and what did you do and are doing? There is a comprehensive text.

You cannot tell how I enjoyed your little visit here. It did me much good. If Harvey is still with you, pray remember me very kindly to him.

. . . I am working most steadily at my Abstract, but it grows to an inordinate length; yet fully to make my view clear (and never giving briefly more than a fact or two, and slurring over difficulties), I cannot make it shorter. It will yet take me three or four months; so slow do I work, though never idle. You cannot imagine what a service you have done me in making me make this Abstract; for though I thought I had got all clear, it has clarified my brains very much, by making me weigh the relative importance of the several elements.

I have been reading with much interest your (as I believe it to be) capital memoir of R. Brown in the *Gardeners' Chronicle.* . . .

C. Darwin to J. D. Hooker.

Down, Oct. 12th, 1858.

. . . I have sent eight copies * by post to Wallace, and will keep the others for him, for I could not think of any one to send any to.

I pray you not to pronounce too strongly against Natural Selection, till you have read my Abstract, for though I dare say you will strike out *many* difficulties, which have never occurred to me; yet you cannot have thought so fully on the subject as I have.

I expect my Abstract will run into a small volume, which will have to be published separately. . . .

What a splendid lot of work you have in hand.

Ever yours,

C. DARWIN.

C. Darwin to J. D. Hooker.

Down, Oct. 13th, 1858.

. . . I have been a little vexed at myself at having asked you not "to pronounce too strongly against Natural Selection." I am sorry to have bothered you, though I have been much interested by your note in answer. I wrote the sentence without reflection. But the truth is, that I have so accustomed myself, partly from being quizzed by my non-naturalist relations, to expect opposition and even contempt, that I forgot for

* Of the joint paper by C. Darwin and A. R. Wallace.

the moment that you are the one living soul from whom I have constantly received sympathy. Believe [me] that I never forget for even a minute how much assistance I have received from you. You are quite correct that I never even suspected that my speculations were a "jam-pot" to you ; indeed, I thought, until quite lately, that my MS. had produced no effect on you, and this has often staggered me. Nor did I know that you had spoken in general terms about my work to our friends, excepting to dear old Falconer, who some few years ago once told me that I should do more mischief than any ten other naturalists would do good, [and] that I had half-spoiled you already ! All this is stupid egotistical stuff, and I write it only because you may think me ungrateful for not having valued and understood your sympathy ; which God knows is not the case. It is an accursed evil to a man to become so absorbed in any subject as I am in mine.

I was in London yesterday for a few hours with Falconer, and he gave me a magnificent lecture on the age of man. We are not upstarts ; we can boast of a pedigree going far back in time coeval with extinct species. He has a grand fact of some large molar tooth in the Trias.

I am quite knocked up, and am going next Monday to revive under Water-cure at Moor Park.

My dear Hooker, yours affectionately,

C. DARWIN.

C. Darwin to J. D. Hooker.

Nov. 1858.

. . . . I had vowed not to mention my everlasting Abstract to you again, for I am sure I have bothered you far more than enough about it ; but, as you allude to its publication, I may say that I have the chapters on Instinct and Hybridism to abstract, which may take a fortnight each ; and my materials for Palæontology, Geographical Distribution, and Affinities, being less worked up, I dare say each of these will take me three weeks, so that I shall not have done at soonest till April, and then my Abstract will in bulk make a small volume. I never give more than one or two instances, and I pass over briefly all difficulties, and yet I cannot make my Abstract shorter, to be satisfactory, than I am now doing, and yet it will expand to a small volume. . . .

[About this time my father revived his old knowledge of beetles in helping his boys in their collecting. He sent a short notice to the 'Entomologist's Weekly Intelligencer,' June 25th, 1859, recording the capture of *Licinus silphoides, Clytus mysticus, Panagæus 4-pustulatus*. The notice begins with the words, "We three very young collectors having lately taken in the parish of Down," &c., and is signed by three of his boys, but was clearly not written by them. I have a vivid recollection of the pleasure of turning out my bottle of dead beetles for my father to name, and the excitement, in which he fully shared, when any of them proved to be uncommon ones. The following letters to Mr. Fox (November 13, 1858), and to Sir John Lubbock, illustrate this point :]

C. Darwin to W. D. Fox.

Down, Nov. 13th [1858].

. . . W., my son, is now at Christ's College, in the rooms above yours. My old Gyp, Impey, was astounded to hear that he was my son, and very simply asked, "Why, has he been long married?" What pleasant hours those were when I used to come and drink coffee with you daily! I am reminded of old days by my third boy having just begun collecting beetles, and he caught the other day *Brachinus crepitans*, of immortal Whittlesea Mere memory. My blood boiled with old ardour when he caught a Licinus—a prize unknown to me . . .

C. Darwin to John Lubbock.

Thursday [before 1857].

DEAR LUBBOCK,—I do not know whether you care about beetles, but for the chance I send this in a bottle, which I never remember having seen ; though it is excessively rash to speak from a twenty-five-year old remembrance. Whenever we meet you can tell me whether you know it. . . .

I feel like an old war-horse at the sound of the trumpet, when I read about the capturing of rare beetles—is not this a magnanimous simile for a decayed entomologist?—It really almost makes me long to begin collecting again. Adios.

"Floreat Entomologia"!—to which toast at Cambridge I have drunk many a glass of wine. So again, "Floreat Entomologia." N.B. I have *not* now been drinking any glasses full of wine.

Yours,

C. D.

C. Darwin to Herbert Spencer.

Down, Nov. 25th [1858].

DEAR SIR,—I beg permission to thank you sincerely for your very kind present of your Essays.* I have already read several of them with much interest. Your remarks on the general argument of the so-called development theory seems to me admirable. I am at present preparing an Abstract of a larger work on the changes of species ; but I treat the subject simply as a naturalist, and not from a general point of view, otherwise, in my opinion, your argument could not have been improved on, and might have been quoted by me with great advantage. Your article on Music has also interested me much, for I had often thought on the subject, and had come

* 'Essays, Scientific, Political, and Speculative,' by Herbert Spencer, 1858–74.

to nearly the same conclusion with you, though unable to support the notion in any detail. Furthermore, by a curious coincidence, expression has been for years a persistent subject with me for *loose* speculation, and I must entirely agree with you that all expression has some biological meaning. I hope to profit by your criticism on style, and with very best thanks, I beg leave to remain, dear Sir,

Yours truly obliged,

C. DARWIN.

C. Darwin to J. D. Hooker.

Down, Dec. 24th [1858].

MY DEAR HOOKER,—Your news about your unsolicited salary and house is jolly, and creditable to the Government. My room (28 × 19), with divided room above, with *all fixtures* (and painted), not furniture, and plastered outside, cost about £500. I am heartily glad of this news.

Your facts about distribution are, indeed, very striking. I remember well that none of your many wonderful facts in your several works, perplexed me, for years, more than the migration having been mainly from north to south, and not in the reverse direction. I have now at last satisfied *myself* (but that is very different from satisfying others) on this head ; but it would take a little volume to fully explain myself. I did not for long see the bearing of a conclusion, at which I had arrived, with respect to this subject. It is, that species inhabiting a very large area, and therefore exist-ing in large numbers, and which have been subjected to the severest competition with many other forms, will have arrived, through natural selection, at a higher stage of per-fection than the inhabitants of a small area. Thus I ex-plain the fact of so many anomalies, or what may be called " living fossils," inhabiting now only fresh water, having been beaten out, and exterminated in the sea, by more improved forms ; thus all existing Ganoid fishes are fresh water, as [are] Lepidosiren and Ornithorhynchus, &c. The plants of Europe with Asia, as being the largest territory, I look at as the most "improved," and therefore as being able to withstand the less-perfected Australian plants ; though these could not resist the Indian. See how all the productions of New Zealand yield to those of Europe. I dare say you will think all this utter bosh, but I believe it to be solid truth.

You will, I think, admit that Australian plants, flourishing so in India, is no argument that they could hold their own against the ten thousand natural contingencies of other plants, insects, animals, &c. &c. With respect to South-West Australia and the Cape, I am shut up, and can only d—n the whole case.

. . . You say you should like to see my MS., but you did read and approved of my long Glacial chapter, and I have not yet written my Abstract on the whole of the Geographical Distribution, nor shall I begin it for two or three weeks. But either Abstract or the old MS. I should be *delighted* to send you, especially the Abstract chapter. . . .

I have now written 330 folio pages of my Abstract, and it will require 150–200 ; so that it will make a printed volume of 400 pages, and must be printed separately, which I think will be better in many respects. The subject really seems to me too large for discussion at any Society, and I believe religion would be brought in by men whom I know.

I am thinking of a 12mo. volume, like Lyell's fourth or fifth edition of the 'Principles.' . . .

I have written you a scandalously long note. So now good bye, my dear Hooker,

Ever yours,
C. DARWIN.

C. Darwin to J. D. Hooker.

Down, Jan. 20th, 1859.

MY DEAR HOOKER,—I should very much like to borrow Heer at some future time, for I want to read nothing perplexing at present till my Abstract is done. Your last very instructive letter shall make me very cautious on the hyperspeculative points we have been discussing.

When you say you cannot master the train of thoughts, I know well enough that they are too doubtful and obscure to be mastered. I have often experienced what you call the humiliating feeling of getting more and more involved in doubt, the more one thinks of the facts and reasoning on doubtful points. But I always comfort myself with thinking of the future, and in the full belief that the problems which we are just entering on, will some day be solved; and if we just break the ground we shall have done some service, even if we reap no harvest.

I quite agree that we only differ in *degree* about the means of dispersal, and that I think a satisfactory amount of accordance. You put in a very striking manner the mutation of our continents, and I quite agree; I doubt only about our oceans.

I also agree (I am in a very agreeing frame of mind) with your *argumentum ad hominem*, about the highness of the Australian Flora from the number of species and genera; but here comes in a superlative bothering element of doubt, viz. the effects of isolation.

The only point in which I *presumptuously* rather demur is about the status of the naturalised plants in Australia. I think Müller speaks of their having spread largely beyond cultivated ground; and I can hardly believe that our European plants would occupy stations so barren that the native plants could not live there. I should require much evidence to make me believe this. I have written this note merely to thank you, as you will see it requires no answer.

I have heard to my amazement this morning from Phillips
that the Geological Council have given me the Wollaston
Medal ! ! !

> Ever yours,
> C. DARWIN.

C. Darwin to J. D. Hooker.

Down, Jan. 23rd, 1859.

. . . I enclose letters to you and me from Wallace. I admire extremely the spirit in which they are written. I never felt
very sure what he would say. He must be an amiable man.
Please return that to me, and Lyell ought to be told how
well satisfied he is. These letters have vividly brought before
me how much I owe to your and Lyell's most kind and
generous conduct in all this affair.

. . . How glad I shall be when the Abstract is finished,
and I can rest ! . . .

C. Darwin to A. R. Wallace.

Down, Jan. 25th [1859].

MY DEAR SIR,—I was extremely much pleased at receiving
three days ago your letter to me and that to Dr. Hooker.
Permit me to say how heartily I admire the spirit in which
they are written. Though I had absolutely nothing whatever
to do in leading Lyell and Hooker to what they thought a
fair course of action, yet I naturally could not but feel anxious
to hear what your impression would be. I owe indirectly
much to you and them ; for I almost think that Lyell would
have proved right, and I should never have completed my
larger work, for I have found my Abstract hard enough with
my poor health, but now, thank God, I am in my last chapter
but one. My Abstract will make a small volume of 400 or
500 pages. Whenever published, I will, of course, send you a
copy, and then you will see what I mean about the part
which I believe selection has played with domestic produc-

tions. It is a very different part, as you suppose, from that played by " Natural Selection." I sent off, by the same address as this note, a copy of the ' Journal of the Linnean Society,' and subsequently I have sent some half-dozen copies of the paper. I have many other copies at your disposal. . . .

I am glad to hear that you have been attending to birds' nests. I have done so, though almost exclusively under one point of view, viz. to show that instincts vary, so that selection could work on and improve them. Few other instincts, so to speak, can be preserved in a Museum.

Many thanks for your offer to look after horses' stripes ; if there are any donkeys, pray add them. I am delighted to hear that you have collected bees' combs. . . . . This is an especial hobby of mine, and I think I can throw a light on the subject. If you can collect duplicates, at no very great expense, I should be glad of some specimens for myself with some bees of each kind. Young, growing, and irregular combs, and those which have not had pupæ, are most valuable for measurements and examination. Their edges should be well protected against abrasion.

Every one whom I have seen has thought your paper very well written and interesting. It puts my extracts (written in 1839, now just twenty years ago !), which I must say in apology were never for an instant intended for publication, into the shade.

You ask about Lyell's frame of mind. I think he is somewhat staggered, but does not give in, and speaks with horror, often to me, of what a thing it would be, and what a job it would be for the next edition of ' The Principles,' if he were "perverted." But he is most candid and honest, and I think will end by being perverted. Dr. Hooker has become almost as heterodox as you or I, and I look at Hooker as *by far* the most capable judge in Europe.

Most cordially do I wish you health and entire success in all your pursuits, and, God knows, if admirable zeal and energy deserve success, most amply do you deserve it. I look at my own career as nearly run out. If I can publish my Abstract and perhaps my greater work on the same subject, I shall look at my course as done.

Believe me, my dear sir, yours very sincerely,

C. DARWIN.

C. Darwin to J. D. Hooker.

Down, March 2nd [1859].

MY DEAR HOOKER,—Here is an odd, though very little, fact. I think it would be hardly possible to name a bird which apparently could have less to do with distribution than a Petrel. Sir W. Milner, at St. Kilda, cut open some young nestling Petrels, and he found large, curious nuts in their crops; I suspect picked up by parent birds from the Gulf stream. He seems to value these nuts excessively. I have asked him (but I doubt whether he will) to send a nut to Sir William Hooker (I gave this address for grandeur's sake) to see if any of you can name it and its native country. Will you *please mention* this to Sir William Hooker, and if the nut does arrive, will you oblige me by returning it to "Sir W. Milner, Bart., Nunappleton, Tadcaster," in a registered letter, and I will repay you postage. Enclose slip of paper with the name and country if you can, and let me hereafter know. Forgive me asking you to take this much trouble; for it is a funny little fact after my own heart.

Now for another subject. I have finished my Abstract of the chapter on Geographical Distribution, as bearing on my subject. I should like you much to read it; but I say this, believing that you will not do so, if, as I believe to be the case, you are extra busy. On my honour, I shall not be mortified, and I earnestly beg you not to do it, if it will bother you. I want it, because I here feel especially unsafe, and errors may have crept in. Also, I should much like to know what parts you will *most vehemently* object to. I know

L 2

we do, and must, differ widely on several heads. Lastly, I should like particularly to know whether I have taken anything from you, which you would like to retain for first publication ; but I think I have chiefly taken from your published works, and, though I have several times, in this chapter and elsewhere, acknowledged your assistance, I am aware that it is not possible for me in the Abstract to do it sufficiently.* But again let me say that you must not offer to read it if very irksome.   It is long—about ninety pages, I expect, when fully copied out.

I hope you are all well. Moor Park has done me some good.

Yours affectionately,

C. DARWIN.

P.S.—Heaven forgive me, here is another question : How far am I right in supposing that with plants, the most important characters for main divisions are embryological ? The seed itself cannot be considered as such, I suppose, nor the albumen, &c.   But I suppose the cotyledons and their position, and the position of the plumule and the radicle, and the position and form of the whole embryo in the seed are embryological, and how far are these very important ? I wish to instance plants as a case of high importance of embryological characters in classification. In the Animal Kingdom there is, of course, no doubt of this.

C. Darwin to J. D. Hooker.

Down, March 5th [1859].

MY DEAR HOOKER,—Many thanks about the seed . . . it is curious. Petrels at St. Kilda apparently being fed by

* "I never did pick any one's pocket, but whilst writing my present chapter I keep on feeling (even when differing most from you) just as if I were stealing from you, so much do I owe to your writings and conversation, so much more than mere acknowledgments show."— Letter to Sir J. D. Hooker, 1859.

seeds raised in the West Indies. It should be noted whether it is a nut ever imported into England. I am *very* glad you will read my Geographical MS. ; it is now copying, and it will (I presume) take ten days or so in being finished ; it shall be sent as soon as done. . . .

I shall be very glad to see your embryological ideas on plants ; by the sentence which I sent you, you will see that I only want one sentence ; if facts are at all, as I suppose, and I shall see this from your note, for sending which very many thanks.

I have been so poorly, the last three days, that I sometimes doubt whether I shall ever get my little volume done, though so nearly completed. . . .

C. Darwin to J. D. Hooker.

Down, March 15th [1859].

MY DEAR HOOKER,—I am *pleased* at what you say of my chapter. You have not attacked it nearly so much as I feared you would. You do not seem to have detected *many* errors. It was nearly all written from memory, and hence I was particularly fearful ; it would have been better if the whole had first been carefully written out, and abstracted afterwards. I look at it as morally certain that it must include much error in some of its general views. I will just run over a few points in your note, but do not trouble yourself to reply without you have something important to say. . . .

. . . I should like to know whether the case of endemic bats in islands struck you ; it has me especially ; perhaps too strongly.

With hearty thanks, ever yours,

C. DARWIN.

P.S.—You cannot tell what a relief it has been to me your looking over this chapter, as I felt very shaky on it.

I shall to-morrow finish my last chapter (except a recapitulation) on Affinities, Homologies, Embryology, &c., and the facts seem to me to come out *very* strong for mutability of species.

I have been much interested in working out the chapter.

I shall now, thank God, begin looking over old first chapters for press.

But my health is now so very poor, that even this will take me long.

C. Darwin to W. D. Fox.

Down, [March] 24th [1859].

MY DEAR FOX,—It was very good of you to write to me in the midst of all your troubles, though you seem to have got over some of them, in the recovery of your wife's and your own health. I had not heard lately of your mother's health, and am sorry to hear so poor an account. But as she does not suffer much, that is the great thing; for mere life I do not think is much valued by the old. What a time you must have had of it, when you had to go backwards and forwards.

We are all pretty well, and our eldest daughter is improving. I can see daylight through my work, and am now finally correcting my chapters for the press; and I hope in a month or six weeks to have proof-sheets. I am weary of my work. It is a very odd thing that I have no sensation that I over-work my brain; but facts compel me to conclude that my brain was never formed for much thinking. We are resolved to go for two or three months, when I have finished, to Ilkley, or some such place, to see if I can anyhow give my health a good start, for it certainly has been wretched of late, and has incapacitated me for everything. You do me injustice when you think that I work for fame; I value it to a certain extent; but, if I know myself, I work from a sort of instinct to try to make out truth. How glad I should be if you could sometime come to Down; especially when I get a little better, as I still hope to be. We have set up a billiard table, and I find it does me a deal of good, and drives the horrid species out of my head. Farewell, my dear old friend.

Yours affectionately,

C. DARWIN.

C. Darwin to C. Lyell.

Down, March 28th [1859].

MY DEAR LYELL,—If I keep decently well, I hope to be able to go to press with my volume early in May. This being so, I want much to beg a little advice from you. From an expression in Lady Lyell's note, I fancy that you have spoken to Murray. Is it so? And is he willing to publish my Abstract? If you will tell me whether anything, and what has passed, I will then write to him. Does he know at all of the subject of the book? Secondly, can you advise me, whether I had better state what terms of publication I should prefer, or first ask him to propose terms? And what do you think would be fair terms for an edition? Share profits, or what?

Lastly, will you be so very kind as to look at the enclosed title and give me your opinion and any criticisms; you must remember that, if I have health and it appears worth doing, I have a much larger and full book on the same subject nearly ready.

My Abstract will be about five hundred pages of the size of your first edition of the 'Elements of Geology.'

Pray forgive me troubling you with the above queries; and you shall have no more trouble on the subject. I hope the world goes well with you, and that you are getting on with your various works.

I am working very hard for me, and long to finish and be free and try to recover some health.

My dear Lyell, ever yours,

C. DARWIN.

Very sincere thanks to you for standing my proxy for the Wollaston Medal.

P.S.—Would you advise me to tell Murray that my book is not more *un*-orthodox than the subject makes inevitable. That I do not discuss the origin of man. That I do not bring in any discussion about Genesis, &c. &c., and only give facts, and such conclusions from them as seem to me fair.

Or had I better say *nothing* to Murray, and assume that he cannot object to this much unorthodoxy, which in fact is not more than any Geological Treatise which runs slap counter to Genesis.

Enclosure.

AN ABSTRACT OF AN ESSAY

ON THE

ORIGIN

OF

SPECIES AND VARIETIES

THROUGH NATURAL SELECTION

BY

CHARLES DARWIN, M.A.

FELLOW OF THE ROYAL, GEOLOGICAL, AND LINNEAN SOCIETIES

LONDON :

&c. &c. &c. &c.

1859.

C. Darwin to C. Lyell.

Down, March 30th [1859].

MY DEAR LYELL,—You have been uncommonly kind in all you have done. You not only have saved me much trouble and some anxiety, but have done all incomparably better than I could have done it. I am much pleased at all you say about Murray. I will write either to-day or to-morrow to him, and will send shortly a large bundle of

MS., but unfortunately I cannot for a week, as the first three chapters are in the copyists' hands.

I am sorry about Murray objecting to the term Abstract, as I look at it as the only possible apology for *not* giving references and facts in full, but I will defer to him and you. I am also sorry about the term "natural selection." I hope to retain it with explanation somewhat as thus :—

"Through natural selection, or the preservation of favoured races."

Why I like the term is that it is constantly used in all works on breeding, and I am surprised that it is not familiar to Murray ; but I have so long studied such works that I have ceased to be a competent judge.

I again most truly and cordially thank you for your really valuable assistance.

Yours most truly,
C. DARWIN.

C. Darwin to J. D. Hooker.

Down, April 2nd [1859].

. . . . I wrote to him [Mr. Murray] and gave him the headings of the chapters, and told him he could not have the MS. for ten days or so ; and this morning I received a letter, offering me handsome terms, and agreeing to publish without seeing the MS. ! So he is eager enough ; I think I should have been cautious, anyhow, but, owing to your letter, I told him most *explicitly* that I accept his offer solely on condition that, after he has seen part or all the MS., he has full power of retracting. You will think me presumptuous, but I think my book will be popular to a certain extent (enough to ensure [against] heavy loss) amongst scientific and semi-scientific men ; why I think so is, because I have found in conversation so great and surprising an interest amongst such men, and some 0-scientific [non-scientific] men on this subject, and all my chapters are not *nearly* so dry and dull as that which you have read on geographical distribution.   Anyhow, Murray ought to be the best judge, and if he chooses to publish it, I think I may wash my hands of all responsibility.   I am sure my friends, *i.e.* Lyell and you, have been *extraordinarily* kind in troubling yourselves on the matter.

I shall be delighted to see you the day before Good Friday ; there would be one advantage for you in any other day—as I believe both my boys come home on that day— and it would be almost impossible that I could send the carriage for you.   There will, I believe, be some relations in the house—but I hope you will not care for that, as we shall easily get as much talking as my *imbecile state* allows.   I shall deeply enjoy seeing you.

. . . . I am tired, so no more.

My dear Hooker, your affectionate,

C. DARWIN.

P.S.—Please to send, well *tied up* with strong string, my Geographical MS., towards the latter half of next week— *i.e.* 7th or 8th—that I may send it with more to Murray ; and God help him if he tries to read it.

. . . . I cannot help a little doubting whether Lyell would take much pains to induce Murray to publish my book ; this was not done at my request, and it rather grates against my pride.

I know that Lyell has been *infinitely* kind about my affair, but your dashed [*i.e.* underlined] "*induce*" gives the idea that Lyell had unfairly urged Murray.

C. Darwin to Asa Gray.

April 4th [1859].

. . . . You ask to see my sheets as printed off; I assure you that it will be the *highest* satisfaction to me to do so : I look at the request as a high compliment.   I shall not, you may depend, forget a request which I look at as a favour. But (and it is a heavy "but" to me) it will be long before I go to press; I can truly say I am *never* idle; indeed, I work too hard for my much weakened health; yet I can do only three hours of work daily, and I cannot at all see when I shall have finished : I have done eleven long chapters, but I have got some other very difficult ones : as palæontology, classifications, and embryology, &c., and I have to correct and add largely to all those done.   I find, alas! each chapter takes me on an average three months, so slow I am.   There is no end to the necessary digressions.   I have just finished a chapter on instinct, and here I found grappling with such a subject as bees' cells, and comparing all my notes made during twenty years, took up a despairing length of time.

But I am running on about myself in a most egotistical style.   Yet I must just say how useful I have again and again found your letters, which I have lately been looking over and quoting! but you need not fear that I shall quote anything you would dislike, for I try to be very cautious on this head.   I most heartily hope you may succeed in getting your " incubus " of old work off your hands, and be in some degree a free man. . . . .

Again let me say that I do indeed feel grateful to you . . .

C. *Darwin to J. Murray.*

Down, April 5th [1859].

MY DEAR SIR,—I send by this post, the Title (with some remarks on a separate page), and the first three chapters. If you have patience to read all Chapter I., I honestly think you will have a fair notion of the interest of the whole book. It may be conceit, but I believe the subject will interest the public, and I am sure that the views are original.   If you think otherwise, I must repeat my request that you will freely reject my work; and though I shall be a little disappointed, I shall be in no way injured.

If you choose to read Chapters II. and III., you will have a dull and rather abstruse chapter, and a plain and interesting one, in my opinion.

As soon as you have done with the MS., please to send it by *careful messenger, and plainly directed,* to Miss G. Tollett, 14, Queen Anne Street, Cavendish Square.

This lady, being an excellent judge of style, is going to look out for errors for me.

You must take your own time, but the sooner you finish, the sooner she will, and the sooner I shall get to press, which I so earnestly wish.

I presume you will wish to see Chapter IV., the key-stone of my arch, and Chapters X. and XI., but please to inform me on this head.

My dear Sir, yours sincerely,

C. DARWIN.

C. Darwin to J. D. Hooker.

Down, April 11th [1859].

. . . I write one line to say that I heard from Murray yesterday, and he says he has read the first three chapters of one MS. (and this includes a very dull one, and he abides by his offer). Hence he does not want more MS., and you can send my Geographical chapter when it pleases you. . . .

[Part of the MS. seems to have been lost on its way back to my father, he wrote (April 14) to Sir J. D. Hooker:

"I have the old MS., otherwise, the loss would have killed me! The worst is now that it will cause delay in getting to press, and *far worst* of all, I lose all advantage of your having looked over my chapter, except the third part returned. I am very sorry Mrs. Hooker took the trouble of copying the two pages."]

C. Darwin to J. D. Hooker.

[April or May, 1859.]

. . . Please do not say to any one that I thought my book on Species would be fairly popular, and have a fairly remunerative sale (which was the height of my ambition), for if it prove a dead failure, it would make me the more ridiculous.

I enclose a criticism, a taste of the future—

*Rev. S. Haughton's Address to the Geological Society, Dublin.**

"This speculation of Messrs. Darwin and Wallace would not be worthy of notice were it not for the weight of authority of the names (*i.e.* Lyell's and yours), under whose auspices it has been brought forward.   If it means what it says, it is a truism ; if it means anything more, it is contrary to fact."

Q. E. D.

C. Darwin to J. D. Hooker.

Down, May 11th [1859].

MY DEAR HOOKER,—Thank you for telling me about obscurity of style.   But on my life no nigger with lash over him could have worked harder at clearness than I have done. But the very difficulty to me, of itself leads to the probability that I fail.   Yet one lady who has read all my MS. has found only two or three obscure sentences, but Mrs. Hooker having so found it, makes me tremble.   I will do my best in proofs.   You are a good man to take the trouble to write about it.

With respect to our mutual muddle,† I never for a moment

---

* Feb. 9, 1858.

† "When I go over the chapter I will see what I can do, but I hardly know how I am obscure, and I think we are somehow in a mutual muddle with respect to each other, from starting from some fundamentally different notions."— Letter of May 6, 1859.

thought we could not make our ideas clear to each other by talk, or if either of us had time to write in extenso.

I imagine from some expressions (but if you ask me what, I could not answer) that you look at variability as some necessary contingency with organisms, and further that there is some necessary tendency in the variability to go on diverging in character or degree. *If you do*, I do not agree. "Reversion" again (a form of inheritance), I look at as in no way directly connected with Variation, though of course inheritance is of fundamental importance to us, for if a variation be not inherited, it is of no signification to us. It was on such points as these *I fancied* that we perhaps started differently.

I fear that my book will not deserve at all the pleasant things you say about it; and Good Lord, how I do long to have done with it!

Since the above was written, I have received and have been *much interested* by A. Gray. I am delighted at his note about my and Wallace's paper. He will go round, for it is futile to give up very many species, and stop at an arbitrary line at others. It is what my grandfather called Unitarianism, "a feather bed to catch a falling Christian." ...

C. Darwin to J. D. Hooker.

Down, May 18th [1859].

MY DEAR HOOKER,—My health has quite failed. I am off to-morrow for a week of Hydropathy. I am very very sorry to say that I cannot look over any proofs * in the week, as my object is to drive the subject out of my head. I shall return to-morrow week. If it be worth while, which probably it is not, you could keep back any proofs till my return home.

In haste, ever yours,

C. DARWIN.

* Of Sir J. D. Hooker's Introduction to the ' Flora of Australia.'

[Ten days later he wrote to Sir J. D. Hooker :

" . . . I write one word to say that I shall return on Saturday, and if you have any proof-sheets to send, I shall be glad to do my best in any criticisms.

I had . . . great prostration of mind and body, but entire rest, and the douche, and 'Adam Bede,' have together done me a world of good."]

C. Darwin to J. Murray.

Down, June 14th [1859].

MY DEAR SIR,—The diagram will do very well, and I will send it shortly to Mr. West to have a few trifling corrections made.

I get on very slowly with proofs. I remember writing to you that I thought there would be not much correction. I honestly wrote what I thought, but was most grievously mistaken. I find the style incredibly bad, and most difficult to make clear and smooth. I am extremely sorry to say, on account of expense, and loss of time for me, that the corrections are very heavy, as heavy as possible. But from casual glances, I still hope that later chapters are not so badly written. How I could have written so badly is quite inconceivable, but I suppose it was owing to my whole attention being fixed on the general line of argument, and not on details. All I can say is, that I am very sorry.

Yours very sincerely,

C. DARWIN.

P.S.—I have been looking at the corrections, and considering them. It seems to me that I shall put you to a quite unfair expense. If you please I should like to enter into some such arrangement as the following : When work completed, you to allow in the account a fairly moderately heavy charge for corrections, and all excess over that to be deducted from my profits, or paid by me individually.

C. Darwin to C. Lyell.

Down, June 21st [1859].

. . . I am working very hard, but get on slowly, for I find that my corrections are terrifically heavy, and the work most difficult to me.   I have corrected 130 pages, and the volume will be about 500.   I have tried my best to make it clear and striking, but very much fear that I have failed—so many discussions are and must be very perplexing.   I have done my best.   If you had all my materials, I am sure you would have made a splendid book.   I long to finish, for I am nearly worn out.

My dear Lyell, ever yours most truly,

C. DARWIN.

C. Darwin to J. D. Hooker.

Down, 22nd [June, 1859].

MY DEAR HOOKER,—I did not answer your pleasant note, with a good deal of news to me, of May 30th, as I have been expecting proofs from you.   But now, having nothing particular to do, I will fly a note, though I have nothing particular to say or ask.   Indeed, how can a man have anything to say, who spends every day in correcting accursed proofs ; and such proofs !   I have fairly to blacken them, and fasten slips of paper on, so miserable have I found the style. You say that you dreamt that my book was *entertaining;* that dream is pretty well over with me, and I begin to fear that the public will find it intolerably dry and perplexing.   But I will never give up that a better man could have made a splendid book out of the materials.   I was glad to hear about Prestwich's paper.*   My doubt has been (and I see Wright

---

* Mr. Prestwich wrote on the occurrence of flint instruments associated with the remains of extinct animals in France.—Proc. R. Soc., 1859.

has inserted the same in the 'Athenæum') whether the pieces of flint are really tools; their numbers make me doubt, and when I formerly looked at Boucher de Perthe's drawings, I came to the conclusion that they were angular fragments broken by ice action.

Did crossing the Acacia do any good? I am so hard worked, that I can make no experiments. I have got only to 150 pages in first proof.

Adios, my dear Hooker, ever yours,

C. DARWIN.

C. Darwin to J. Murray.

Down, July 25th [1859].

MY DEAR SIR,—I write to say that five sheets are returned to the printers ready to strike off, and two more sheets require only a revise; so that I presume you will soon have to decide what number of copies to print off.

I am quite incapable of forming any opinion. I think I have got the style *fairly* good and clear, with infinite trouble. But whether the book will be successful to a degree to satisfy you, I really cannot conjecture. I heartily hope it may.

My dear Sir, yours very sincerely,

C. DARWIN.

C. Darwin to A. R. Wallace.

Down, Aug. 9th, 1859.

MY DEAR MR. WALLACE,—I received your letter and memoir * on the 7th, and will forward it to-morrow to the Linnean Society. But you will be aware that there is no meeting till the beginning of November. Your paper seems to me *admirable* in matter, style, and reasoning; and I thank

---

* This seems to refer to Mr. Wallace's paper, "On the Zoological Geography of the Malay Archipelago," 'Linn. Soc. Journ.,' 1860.

you for allowing me to read it.   Had I read it some months
ago, I should have profited by it for my forthcoming volume.
But my two chapters on this subject are in type, and, though
not yet corrected, I am so wearied out and weak in health,
that I am fully resolved not to add one word, and merely
improve the style.   So you will see that my views are nearly
the same with yours, and you may rely on it that not one
word shall be altered owing to my having read your ideas.
Are you aware that Mr. W. Earl * [sic] published several years
ago the view of distribution of animals in the Malay Archipelago, in relation to the depth of the sea between the islands ?
I was much struck with this, and have been in the habit of
noting all facts in distribution in that archipelago, and elsewhere, in this relation.   I have been led to conclude that
there has been a good deal of naturalisation in the different
Malay islands, and which I have thought, to a certain extent,
would account for anomalies.   Timor has been my greatest
puzzle.   What do you say to the peculiar Felis there?   I
wish that you had visited Timor; it has been asserted that a
fossil mastodon's or elephant's tooth (I forget which) has been
found there, which would be a grand fact.   I was aware that
Celebes was very peculiar; but the relation to Africa is quite
new to me, and marvellous, and almost passes belief.   It is as
anomalous as the relation of *plants* in S.W. Australia to the
Cape of Good Hope.   I differ *wholly* from you on the colonisa-
tion of oceanic islands, but you will have *every one* else
on your side.   I quite agree with respect to all islands not
situated far in the ocean.   I quite agree on the little occasional intermigration between lands [islands?] when once
pretty well stocked with inhabitants, but think this does not
apply to rising and ill-stocked islands.   Are you aware that
annually birds are blown to Madeira, the Azores (and to
Bermuda from America)?   I wish I had given a fuller abstract
of my reasons for not believing in Forbes's great continental

* Probably Mr. W. Earle's paper, Geographical Soc. Journal, 1845.

extensions; but it is too late, for I will alter nothing—I am worn out, and must have rest. Owen, I do not doubt, will bitterly oppose us. . . . Hooker is publishing a grand Introduction to the Flora of Australia, and goes the whole length. I have seen proofs of about half. With every good wish.

Believe me, yours very sincerely,

C. DARWIN.

C. Darwin to J. D. Hooker.

Down, Sept. 1st [1859].

. . . I am not surprised at your finding your Introduction very difficult. But do not grudge the labour, and do not say you " have burnt your fingers," and are " deep in the mud"; for I feel sure that the result will be well worth the labour. Unless I am a fool, I must be a judge to some extent of the value of such general essays, and I am fully convinced that yours are the most valuable ever published.

I have corrected all but the last two chapters of my book, and hope to have done revises and all in about three weeks, and then I (or we all) shall start for some months' hydropathy ; my health has been very bad, and I am becoming as weak as a child, and incapable of doing anything whatever, except my three hours daily work at proof-sheets. God knows whether I shall ever be good for anything again, perhaps a long rest and hydropathy may do something.

I have not had A. Gray's Essay, and should not feel up to criticise it, even if I had the impertinence and courage. You will believe me that I speak strictly the truth when I say that your Australian Essay is *extremely* interesting to me, rather too much so. I enjoy reading it over, and if you think my criticisms are worth anything to you, I beg you to send the sheets (if you can give me time for good days); but unless I can render you any little, however little assistance,

I would rather read the essay when published. Pray understand that I should be *truly* vexed not to read them, if you
wish it for your own sake.

I had a terribly long fit of sickness yesterday, which makes
the world rather extra gloomy to-day, and I have an insanely
strong wish to finish my accursed book, such corrections every
page has required as I never saw before. It is so weariful,
killing the whole afternoon, after 12 o'clock doing nothing
whatever. But I will grumble no more. So farewell, we
shall meet in the winter I trust.

Farewell, my dear Hooker, your affectionate friend,

C. DARWIN.

C. Darwin to C. Lyell.

Down, Sept. 2nd [1859].

. . . I am very glad you wish to see my clean sheets : I should
have offered them, but did not know whether it would bore
you ; I wrote by this morning's post to Murray to send them.
Unfortunately I have not got to the part which will interest
you, I think most, and which tells most in favour of the view,
viz. Geological Succession, Geographical Distribution, and especially Morphology, Embryology and Rudimentary Organs. I
will see that the remaining sheets, when printed off, are sent to
you. But would you like for me to send the last and perfect
revises of the sheets as I correct them ? if so, send me your
address in a blank envelope. I hope that you will read all,
whether dull (especially latter part of Chapter II.) or not, for
I am convinced there is not a sentence which has not a
bearing on the whole argument. You will find Chapter IV.
perplexing and unintelligible, without the aid of the enclosed
queer diagram,* of which I send an old and useless proof. I
have, as Murray says, corrected so heavily, as almost to have
re-written it ; but yet I fear it is poorly written. Parts are

* The diagram illustrates descent with divergence.

intricate ; and I do not think that even you could make them
quite clear.   Do not, I beg, be in a hurry in committing
yourself (like so many naturalists) to go a certain length and
no further ; for I am deeply convinced that it is absolutely
necessary to go the whole vast length, or stick to the creation
of each separate species ; I argue this point briefly in the last
chapter.   Remember that your verdict will probably have
more influence than my book in deciding whether such
views as I hold will be admitted or rejected at present ; in
the future I cannot doubt about their admittance, and our
posterity will marvel as much about the current belief as we
do about fossil shells having been thought to have been
created as we now see them.   But forgive me for running
on about my hobby-horse. . . .

C. Darwin to J. D. Hooker.

Down, [Sept.] 11th [1859].

MY DEAR HOOKER,—I corrected the last proof yesterday,
and I have now my revises, index, &c., which will take me
near to the end of the month.   So that the neck of my
work, thank God, is broken.

I write now to say that I am uneasy in my conscience
about hesitating to look over your proofs, but I was feeling
miserably unwell and shattered when I wrote.   I do not
suppose I could be of hardly any use, but if I could, pray
send me any proofs.   I should be (and fear I was) the most
ungrateful man to hesitate to do anything for you after some
fifteen or more years' help from you.

As soon as ever I have fairly finished I shall be off to Ilkley,
or some other Hydropathic establishment.   But I shall be some
time yet, as my proofs have been so utterly obscured with
corrections, that I have to correct heavily on revises.

Murray proposes to publish the first week in November.
Oh, good heavens, the relief to my head and body to banish
the whole subject from my mind !

I hope to God, you do not think me a brute about your proof-sheets.

Farewell, yours affectionately,

C. DARWIN.

C. Darwin to C. Lyell.

Down, Sept. 20th [1859].

MY DEAR LYELL.—You once gave me intense pleasure, or rather delight, by the way you were interested, in a manner I never expected, in my Coral Reef notions, and now you have again given me similar pleasure by the manner you have noticed my species work.* Nothing could be more satisfactory to me, and I thank you for myself, and even more for the subject's sake, as I know well that the sentence will make many fairly consider the subject, instead of ridiculing it. Although your previously felt doubts on the immutability of species, may have more influence in converting you (if you be converted) than my book; yet as I regard your verdict as far more important in my own eyes, and I believe in the eyes of the world than of any other dozen men, I am naturally very anxious about it. Therefore let me beg you to keep your mind open till you receive (in perhaps a fortnight's time) my latter chapters, which are the most

---

* Sir Charles was President of the Geological section at the meeting of the British Association at Aberdeen in 1859. The following passage occurs in the address: " On this difficult and mysterious subject a work will very shortly appear by Mr. Charles Darwin, the result of twenty years of observations and experiments in Zoology, Botany, and Geology, by which he has been led to the conclusion that those powers of nature which give rise to races and permanent varieties in animals and plants, are the same as those which in much longer periods produce species, and in a still longer series of ages give rise to differences of generic rank. He appears to me to have succeeded by his investigations and reasonings in throwing a flood of light on many classes of phenomena connected with the affinities, geographical distribution, and geological succession of organic beings, for which no other hypothesis has been able, or has even attempted to account."

important of all on the favourable side. The last chapter, which sums up, and balances in a mass, all the arguments contra and pro, will, I think, be useful to you. I cannot too strongly express my conviction of the general truth of my doctrines, and God knows I have never shirked a difficulty. I am foolishly anxious for your verdict, not that I shall be disappointed if you are not converted; for I remember the long years it took me to come round; but I shall be most deeply delighted if you do come round, especially if I have a fair share in the conversion, I shall then feel that my career is run, and care little whether I ever am good for anything again in this life.

Thank you much for allowing me to put in the sentence about your grave doubt.* So much and too much about myself.

I have read with extreme interest in the Aberdeen paper about the flint tools; you have made the whole case far clearer to me; I suppose that you did not think the evidence sufficient about the Glacial period.

With cordial thanks for your splendid notice of my book.

Believe me, my dear Lyell, your affectionate disciple,

CHARLES DARWIN.

C. Darwin to W. D. Fox.

Down, Sept. 23rd [1859].

MY DEAR FOX,—I was very glad to get your letter a few days ago. I was wishing to hear about you, but have been in such an absorbed, slavish, overworked state, that I had not heart without compulsion to write to any one or do anything beyond my daily work. Though your account of yourself is better, I cannot think it at all satisfactory, and I wish you would soon go to Malvern again. My father used to believe largely in an old saying that, if a man grew thinner between

---

* As to the immutability of species, ' Origin,' ed. i., p. 310.

fifty and sixty years of age, his chance of long life was poor, and that on the contrary it was a very good sign if he grew fatter; so that your stoutness, I look at as a very good omen. My health has been as bad as it well could be all this summer; and I have kept on my legs, only by going at short intervals to Moor Park; but I have been better lately, and, thank Heaven, I have at last as good as done my book, having only the index and two or three revises to do. It will be published in the first week in November, and a copy shall be sent you. Remember it is only an Abstract (but has cost me above thirteen months to write!!), and facts and authorities are far from given in full. I shall be curious to hear what you think of it, but I am not so silly as to expect to convert you. Lyell has read about half of the volume in clean sheets, and gives me very great *kudos*. He is wavering so much about the immutability of species, that I expect he will come round. Hooker has come round, and will publish his belief soon. So much for my abominable volume, which has cost me so much labour that I almost hate it. On October 3rd I start for Ilkley, but shall take three days for the journey! It is so late that we shall not take a house; but I go there alone for three or four weeks; then return home for a week and go to Moor Park for three or four weeks, and then I shall get a moderate spell of hydropathy; and I intend, if I can keep to my resolution, of being idle this winter. But I fear *ennui* will be as bad as a bad stomach. . . .

C. Darwin to C. Lyell.

Down, Sept. 25th [1859].

MY DEAR LYELL,—I send by this post four corrected sheets. I have altered the sentence about the Eocene fauna being beaten by recent, thanks to your remark. But I imagined that it would have been clear that I supposed the climate to be nearly similar; you do not doubt, I imagine, that the climate

of the Eocene and recent periods in *different* parts of the
world could be matched. Not that I think climate nearly so
important as most naturalists seem to think. In my opinion
no error is more mischievous than this.

I was very glad to find that Hooker, who read over, in
MS., my Geographical chapters, quite agreed in the view of
the greater importance of organic relations. I should like
you to consider p. 77 and reflect on the case of any organism
in the midst of its range.

I shall be curious hereafter to hear what you think of distribution during the glacial and preceding warmer periods.
I am so glad you do not think the Chapter on the Imperfection of the Geological Record exaggerated ; I was more
fearful about this chapter than about any part.

Embryology in Chapter VIII. is one of my strongest points
I think. But I must not bore you by running on. My mind
is so wearisomely full of the subject.

I do thank you for your eulogy at Aberdeen. I have
been so wearied and exhausted of late that I have for months
doubted whether I have not been throwing away time and
labour for nothing. But now I care not what the universal
world says ; I have always found you right, and certainly on
this occasion I am not going to doubt for the first time.
Whether you go far, or but a very short way with me and others
who believe as I do, I am contented, for my work cannot be
in vain. You would laugh if you knew how often I have read
your paragraph, and it has acted like a little dram. . . .

Farewell,

C. DARWIN.

C. Darwin to C. Lyell.

Down, Sept. 30th [1859].

MY DEAR LYELL,—I sent off this morning the last sheets,
but without index, which is not in type. I look at you as my
Lord High Chancellor in Natural Science, and therefore

I request you, after you have finished, just to *re-run* over the heads in the recapitulation-part of last chapter. I shall be deeply anxious to hear what you decide (if you are able to decide) on the balance of the pros and contras given in my volume, and of such other pros and contras as may occur to you. I hope that you will think that I have given the difficulties fairly. I feel an entire conviction that if you are now staggered to any moderate extent, you will come more and more round, the longer you keep the subject at all before your mind. I remember well how many long years it was before I could look into the face of some of the difficulties and not feel quite abashed. I fairly struck my colours before the case of neuter insects.

I suppose that I am a very slow thinker, for you would be surprised at the number of years it took me to see clearly what some of the problems were which had to be solved, such as the necessity of the principle of divergence of character, the extinction of intermediate varieties, on a continuous area, with graduated conditions; the double problem of sterile first crosses and sterile hybrids, &c. &c.

Looking back, I think it was more difficult to see what the problems were than to solve them, so far as I have succeeded in doing, and this seems to me rather curious. Well, good or bad, my work, thank God, is over; and hard work, I can assure you, I have had, and much work which has never borne fruit. You can see, by the way I am scribbling, that I have an idle and rainy afternoon. I was not able to start for Ilkley yesterday as I was too unwell; but I hope to get there on Tuesday or Wednesday. Do, I beg you, when you have finished my book and thought a little over it, let me hear from you. Never mind and pitch into me, if you think it requisite; some future day, in London possibly, you may give me a few criticisms in detail, that is, if you have scribbled any remarks on the margin, for the chance of a second edition.

Murray has printed 1250 copies, which seems to me rather too large an edition, but I hope he will not lose.

I make as much fuss about my book as if it were my first. Forgive me, and believe me, my dear Lyell,

Yours most sincerely,

C. DARWIN.

C. Darwin to J. D. Hooker.

Ilkley, Yorkshire, Oct. 15th [1859].

MY DEAR HOOKER,—Be a good man and screw out time enough to write me a note and tell me a little about yourself, your doings, and belongings.

Is your Introduction fairly finished ? I know you will abuse it, and I know well how much I shall like it. I have been here nearly a fortnight, and it has done me very much good, though I sprained my ankle last Sunday, which has quite stopped walking. All my family come here on Monday to stop three or four weeks, and then I shall go back to the great establishment, and stay a fortnight ; so that if I can keep my spirits, I shall stay eight weeks here, and thus give hydropathy a fair chance. Before starting here I was in an awful state of stomach, strength, temper, and spirits. My book has been completely finished some little time ; as soon as copies are ready, of course one will be sent you. I hope you will mark your copy with scores, so that I may profit by any criticisms. I should like to hear your general impression. From Lyell's letters, he thinks favourably of it, but seems staggered by the lengths to which I go. But if you go any considerable length in the admission of modification, I can see no possible means of drawing the line, and saying here you must stop. Lyell is going to reread my book, and I yet entertain hopes that he will be converted, or perverted, as he calls it. Lyell has been *extremely* kind in writing me three volume-like letters ; but he says nothing about dispersal during the

Glacial period.   I should like to know what he thinks on this head.   I have one question to ask: Would it be any good to send a copy of my book to Decaisne? and do you know any philosophical botanists on the Continent, who read English and care for such subjects? if so, give me their addresses. How about Andersson in Sweden?   You cannot think how refreshing it is to idle away the whole day, and hardly ever think in the least about my confounded book which half-killed me.   I much wish I could hear of your taking a real rest.   I know how very strong you are mentally, but I never will believe you can go on working as you have worked of late with impunity.   You will some day stretch the string too tight.   Farewell, my good, and kind, and dear friend,

Yours affectionately,

C. Darwin.

C. Darwin to T. H. Huxley.

Ilkley, Yorkshire, Oct. 15th [1859].

My dear Huxley, — I am here hydropathising and coming to life again, after having finished my accursed book, which would have been easy work to any one else, but half-killed me.   I have thought you would give me one bit of information, and I know not to whom else to apply ; viz., the addresses of Barrande, Von Siebold, Keyserling (I dare say Sir Roderick would know the latter).

Can you tell me of any good and *speculative* foreigners to whom it would be worth while to send copies of my book, on the ' Origin of Species'?   I doubt whether it is worth sending to Siebold.   I should like to send a few copies about, but how many I can afford I know not yet till I hear what price Murray affixes.

I need not say that I will send, of course, one to you, in the first week of November.   I hope to send copies abroad immediately.   I shall be *intensely* curious to hear what effect

the book produces on you. I know that there will be much
in it which you will object to, and I do not doubt many
errors. I am very far from expecting to convert you to
many of my heresies; but if, on the whole, you and two or
three others think I am on the right road, I shall not care
what the mob of naturalists think. The penultimate chapter,*
though I believe it includes the truth, will, I much fear, make
you savage. Do not act and say, like Macleay versus
Fleming, " I write with aqua fortis to bite into brass."

Ever yours,
C. DARWIN.

C. Darwin to C. Lyell.

Ilkley, Yorkshire.
Oct. 20th [1859].

MY DEAR LYELL,—I have been reading over all your letters consecutively, and I do not feel that I have thanked you
half enough for the extreme pleasure which they have given
me, and for their utility. I see in them evidence of fluctua-
tion in the degree of credence you give to the theory ; nor am
I at all surprised at this, for many and many fluctuations I
have undergone.

There is one point in your letter which I did not notice,
about the animals (and many plants) naturalised in Australia,
which you think could not endure without man's aid. I cannot see how man does aid the feral cattle. But, letting that
pass, you seem to think, that because they suffer prodigious
destruction during droughts, they would all be destroyed. In
the " grandes secos " of La Plata, the indigenous animals, such
as the American deer, die by thousands, and suffer apparently
as much as the cattle. In parts of India, after a drought, it
takes ten or more years before the indigenous mammals get

---

* Chapter XIII. is on Classification, Morphology, Embryology, and
Rudimentary Organs.

up to their full number again. Your argument would, I think, apply to the aborigines as well as to the feral.

An animal or plant which becomes feral in one small territory might be destroyed by climate, but I can hardly believe so, when once feral over several large territories. Again, I feel inclined to swear at climate : do not think me impudent for attacking you about climate. You say you doubt whether man could have existed under the Eocene climate, but man can now withstand the climate of Esquimaux-land and West Equatorial Africa; and surely you do not think the Eocene climate differed from the present throughout all Europe, as much as the Arctic regions differ from Equatorial Africa ?

With respect to organisms being created on the American type in America, it might, I think, be said that they were so created to prevent them being too well created, so as to beat the aborigines ; but this seems to me, somehow, a monstrous doctrine.

I have reflected a good deal on what you say on the necessity of continued intervention of creative power. I cannot see this necessity ; and its admission, I think, would make the theory of Natural Selection valueless. Grant a simple Archetypal creature, like the Mud-fish or Lepidosiren, with the five senses and some vestige of mind, and I believe natural selection will account for the production of every vertebrate animal.

Farewell ; forgive me for indulging in this prose, and believe me, with cordial thanks,

Your ever attached disciple,
C. DARWIN.

P.S.—When, and if, you reread, I supplicate you to write on the margin the word " expand," when too condensed, or " not clear," or " ? ". Such marks would cost you little trouble, and I could copy them and reflect on them, and their value would be infinite to me.

My larger book will have to be wholly re-written, and not merely the present volume expanded ; so that I want to waste as little time over this volume as possible, if another edition be called for ; but I fear the subject will be too perplexing, as I have treated it, for general public.

C. Darwin to J. D. Hooker.

Ilkley, Yorkshire.
Sunday [Oct. 23rd, 1859].

MY DEAR HOOKER,—I congratulate you on your 'Introduction'* being in fact finished.   I am sure from what I read of it (and deeply I shall be interested in reading it straight through), that it must have cost you a prodigious amount of labour and thought.   I shall like very much to see the sheet, which you wish me to look at.   Now I am so completely a gentleman, that I have sometimes a little difficulty to pass the day ; but it is astonishing how idle a three weeks I have passed.   If it is any comfort to you, pray delude yourself by saying that you intend "sticking to humdrum science."   But I believe it just as much as if a plant were to say that, "I have been growing all my life, and, by Jove, I will stop growing." You cannot help yourself ; you are not clever enough for that. You could not even remain idle, as I have done, for three weeks !   What you say about Lyell pleases me exceedingly ; I had not at all inferred from his letters that he had come so much round.   I remember thinking, above a year ago, that if ever I lived to see Lyell, yourself, and Huxley come round, partly by my book, and partly by their own reflections, I should feel that the subject is safe, and all the world might rail, but that ultimately the theory of Natural Selection (though, no doubt, imperfect in its present condition, and embracing many errors) would prevail.   Nothing will ever convince me that three such men, with so much diversified

* 'Australian Flora.'

knowledge, and so well accustomed to search for truth, could err greatly. I have spoken of you here as a convert made by me ; but I know well how much larger the share has been of your own self-thought. I am intensely curious to hear Huxley's opinion of my book. I fear my long discussion on Classification will disgust him ; for it is much opposed to what he once said to me.

But, how I am running on ! You see how idle I am ; but I have so enjoyed your letter that you must forgive me. With respect to migration during the Glacial period : I think Lyell quite comprehends, for he has given me a supporting fact. But, perhaps, he unconsciously hates (do not say so to him) the view, as slightly staggering him on his favourite theory of all changes of climate being due to changes in the relative position of land and water.

I will send copies of my book to all the men specified by you ; . . . would you be so kind as to add title, as Doctor, or Professor, or Monsieur, or Von, and initials (when wanted), and addresses to the names on the enclosed list, and let me have it pretty *soon*, as towards the close of this week Murray says the copies to go abroad will be ready. I am anxious to get my view generally known, and not, I hope and think, for mere personal conceit. . . . .

C. Darwin to C. Lyell.

Ilkley, Yorkshire, Oct. 25th [1859].

. . . Our difference on "principle of improvement" and "power of adaptation" is too profound for discussion by letter. If I am wrong, I am quite blind to my error. If I am right, our difference will be got over only by your re-reading carefully and reflecting on my first four chapters. I supplicate you to read these again carefully. The so-called improvement of our Shorthorn cattle, pigeons, &c., does not presuppose or require any aboriginal "power of adaptation," or "principle of improvement ;" it requires only diversified variability, and man to select or take advantage of those modifications which are useful to him ; so under nature any slight modification which *chances* to arise, and is useful to any creature, is selected or preserved in the struggle for life ; any modification which is injurious is destroyed or rejected ; any which is neither useful nor injurious will be left a fluctuating element. When you contrast natural selection and "improvement," you seem always to overlook (for I do not see how you can deny) that every step in the natural selection of each species implies improvement in that species in relation to its conditions of life. No modification can be selected without it be an improvement or advantage. Improvement implies, I suppose, each form obtaining many parts or organs, all excellently adapted for their functions. As each species is improved, and as the number of forms will have increased, if we look to the whole course of time, the organic condition of life for other forms will become more complex, and there will be a necessity for other forms to become improved, or they will be exterminated ; and I can see no limit to this process of improvement, without the intervention of any other and direct principle of improvement. All this seems to me quite compatible with certain forms fitted for simple conditions, remaining unaltered, or being degraded.

If I have a second edition, I will reiterate " Natural Selection, and as a general consequence, Natural Improvement."

As you go, as far as you do, I begin strongly to think, judging from myself, that you will go much further. How slowly the older geologists admitted your grand views on existing geological causes of change !

If at any time you think I can answer any question, it is a real pleasure to me to write.

Yours affectionately,
C. Darwin.

C. Darwin to J. Murray.

MY DEAR SIR,—I have received your kind note and the copy ; I am infinitely pleased and proud at the appearance of my child.

I quite agree to all you propose about price. But you are really too generous about the, to me, scandalously heavy corrections. Are you not acting unfairly towards yourself ? Would it not be better at least to share the £72 8s. ? I shall be fully satisfied, for I had no business to send, though quite unintentionally and unexpectedly, such badly composed MS. to the printers.

Thank you for your kind offer to distribute the copies to my friends and assisters as soon as possible. Do not trouble yourself much about the foreigners, as Messrs. Williams and Norgate have most kindly offered to do their best, and they are accustomed to send to all parts of the world.

I will pay for my copies whenever you like. I am so glad that you were so good as to undertake the publication of my book.

My dear Sir, yours very sincerely,

CHARLES DARWIN.

P.S.—Please do not forget to let me hear about two days before the copies are distributed.

I do not know when I shall leave this place, certainly not for several weeks. Whenever I am in London I will call on you.

CHAPTER V.

BY PROFESSOR HUXLEY.

ON THE RECEPTION OF THE 'ORIGIN OF SPECIES.'

To the present generation, that is to say, the people a few years on the hither and thither side of thirty, the name of Charles Darwin stands alongside of those of Isaac Newton and Michael Faraday; and, like them, calls up the grand ideal of a searcher after truth and interpreter of Nature. They think of him who bore it as a rare combination of genius, industry, and unswerving veracity, who earned his place among the most famous men of the age by sheer native power, in the teeth of a gale of popular prejudice, and uncheered by a sign of favour or appreciation from the official fountains of honour; as one who, in spite of an acute sensitiveness to praise and blame, and notwithstanding provocations which might have excused any outbreak, kept himself clear of all envy, hatred, and malice, nor dealt otherwise than fairly and justly with the unfairness and injustice which was showered upon him; while, to the end of his days, he was ready to listen with patience and respect to the most insignificant of reasonable objectors.

And with respect to that theory of the origin of the forms of life peopling our globe, with which Darwin's name is bound up as closely as that of Newton with the theory of gravitation, nothing seems to be further from the mind of the present generation than any attempt to smother it with ridicule or to crush it by vehemence of denunciation. "The struggle for

existence," and " Natural selection," have become household
words and every-day conceptions. The reality and the importance of the natural processes on which Darwin founds
his deductions are no more doubted than those of growth
and multiplication ; and, whether the full potency attributed
to them is admitted or not, no one doubts their vast and far-
reaching significance. Wherever the biological sciences are
studied, the ' Origin of Species ' lights the path of the in-
vestigator ; wherever they are taught it permeates the course
of instruction. Nor has the influence of Darwinian ideas
been less profound, beyond the realms of Biology. The
oldest of all philosophies, that of Evolution, was bound hand
and foot and cast into utter darkness during the millennium
of theological scholasticism. But Darwin poured new life-
blood into the ancient frame ; the bonds burst, and the
revivified thought of ancient Greece has proved itself to be
a more adequate expression of the universal order of things
than any of the schemes which have been accepted by the
credulity and welcomed by the superstition of seventy later
generations of men.

To any one who studies the signs of the times, the
emergence of the philosophy of Evolution, in the attitude of
claimant to the throne of the world of thought, from the
limbo of hated and, as many hoped, forgotten things, is the
most portentous event of the nineteenth century. But the
most effective weapons of the modern champions of Evolution
were fabricated by Darwin ; and the ' Origin of Species ' has
enlisted a formidable body of combatants, trained in the severe school of Physical Science, whose ears might have long
remained deaf to the speculations of *a priori* philosophers.

I do not think that any candid or instructed person will
deny the truth of that which has just been asserted. He may
hate the very name of Evolution, and may deny its pretensions
as vehemently as a Jacobite denied those of George the Second.
But there it is—not only as solidly seated as the Hanoverian

dynasty, but happily independent of Parliamentary sanction—
and the dullest antagonists have come to see that they have
to deal with an adversary whose bones are to be broken by
no amount of bad words.

Even the theologians have almost ceased to pit the plain
meaning of Genesis against the no less plain meaning of
Nature. Their more candid, or more cautious, representatives
have given up dealing with Evolution as if it were a damnable
heresy, and have taken refuge in one of two courses. Either
they deny that Genesis [was meant to teach scientific truth,
and thus save the veracity of the record at the expense of its
authority; or they expend their energies in devising the cruel
ingenuities of the reconciler, and torture texts in the vain
hope of making them confess the creed of Science. But when
the *peine forte et dure* is over, the antique sincerity of the venerable sufferer always reasserts itself. Genesis is honest to the
core, and professes to be no more than it is, a repository of
venerable traditions of unknown origin, claiming no scientific
authority and possessing none.

As my pen finishes these passages, I can but be
amused to think what a terrible hubbub would have been
made (in truth was made) about any similar expressions of
opinion a quarter of a century ago. In fact, the contrast
between the present condition of public opinion upon the
Darwinian question; between the estimation in which
Darwin's views are now held in the scientific world; between
the acquiescence, or at least quiescence, of the theologians of
the self-respecting order at the present day and the out-
burst of antagonism on all sides in 1858–9, when the new
theory respecting the origin of species first became known to
the older generation to which I belong, is so startling that,
except for documentary evidence, I should be sometimes
inclined to think my memories dreams. I have a great
respect for the younger generation myself (they can write our
lives, and ravel out all our follies, if they choose to take the

trouble, by and by), and I should be glad to be assured that the feeling is reciprocal; but I am afraid that the story of our dealings with Darwin may prove a great hindrance to that veneration for our wisdom which I should like them to display. We have not even the excuse that, thirty years ago, Mr. Darwin was an obscure novice, who had no claims on our attention. On the contrary, his remarkable zoological and geological investigations had long given him an assured position among the most eminent and original investigators of the day; while his charming 'Voyage of a Naturalist' had justly earned him a wide-spread reputation among the general public. I doubt if there was any man then living who had a better right to expect that anything he might choose to say on such a question as the Origin of Species would be listened to with profound attention, and discussed with respect; and there was certainly no man whose personal character should have afforded a better safeguard against attacks, instinct with malignity and spiced with shameless impertinences.

Yet such was the portion of one of the kindest and truest men that it was ever my good fortune to know; and years had to pass away before misrepresentation, ridicule, and denunciation, ceased to be the most notable constituents of the majority of the multitudinous criticisms of his work which poured from the press. I am loth to rake any of these ancient scandals from their well-deserved oblivion; but I must make good a statement which may seem overcharged to the present generation, and there is no *pièce justificative* more apt for the purpose, or more worthy of such dishonour, than the article in the 'Quarterly Review' for July 1860.* Since Lord Brougham

* I was not aware when I wrote these passages that the authorship of the article had been publicly acknowledged. Confession unaccompanied by penitence, however, affords no ground for mitigation of judgment; and the kindliness with which Mr. Darwin speaks of his assailant, Bishop Wilberforce (Vol. II. pp. 325, 329, 332), is so striking an exemplification of his singular gentleness and modesty, that it rather increases one's indignation against the presumption of his critic.

assailed Dr. Young, the world has seen no such specimen of the insolence of a shallow pretender to a Master in Science as this remarkable production, in which one of the most exact of observers, most cautious of reasoners, and most candid of expositors, of this or any other age, is held up to scorn as a "flighty" person, who endeavours "to prop up his utterly rotten fabric of guess and speculation," and whose "mode of dealing with nature" is reprobated as "utterly dishonourable to Natural Science." And all this high and mighty talk, which would have been indecent in one of Mr. Darwin's equals, proceeds from a writer whose want of intelligence, or of conscience, or of both, is so great, that, by way of an objection to Mr. Darwin's views, he can ask, "Is it credible that all favourable varieties of turnips are tending to become men;" who is so ignorant of paleontology, that he can talk of the "flowers and fruits" of the plants of the carboniferous epoch; of comparative anatomy, that he can gravely affirm the poison apparatus of the venomous snakes to be "entirely separate from the ordinary laws of animal life, and peculiar to themselves;" of the rudiments of physiology, that he can ask, "what advantage of life could alter the shape of the corpuscles into which the blood can be evaporated?" Nor does the reviewer fail to flavour this outpouring of preposterous incapacity with a little stimulation of the *odium theologicum*. Some inkling of the history of the conflicts between Astronomy, Geology, and Theology, leads him to keep a retreat open by the proviso that he cannot "consent to test the truth of Natural Science by the word of Revelation;" but, for all that, he devotes pages to the exposition of his conviction that Mr. Darwin's theory "contradicts the revealed relation of the creation to its Creator," and is "inconsistent with the fulness of his glory."

If I confine my retrospect of the reception of the 'Origin of Species' to a twelvemonth, or thereabouts, from the time of its publication, I do not recollect anything quite so foolish and unmannerly as the 'Quarterly Review' article, unless, perhaps, the address of a Reverend Professor to the Dublin Geological Society might enter into competition with it. But a large proportion of Mr. Darwin's critics had a lamentable resemblance to the 'Quarterly' reviewer, in so far as they lacked either the will, or the wit, to make themselves masters of his doctrine ; hardly any possessed the knowledge required to follow him through the immense range of biological and geological science which the 'Origin' covered ; while, too commonly, they had prejudged the case on theological grounds, and, as seems to be inevitable when this happens, eked out lack of reason by superfluity of railing.

But it will be more pleasant and more profitable to consider those criticisms, which were acknowledged by writers of scientific authority, or which bore internal evidence of the greater or less competency and, often, of the good faith, of their authors. Restricting my survey to a twelvemonth, or thereabouts, after the publication of the 'Origin,' I find among such critics Louis Agassiz ;* Murray, an excellent entomologist ; Harvey, a botanist of considerable repute ; and the author of an article in the 'Edinburgh Review,' all strongly adverse to Darwin. Pictet, the distinguished and widely learned paleontologist of Geneva, treats Mr. Darwin with a respect which forms a grateful contrast to the tone of some of the preceding writers, but consents to go with him

---

* "The arguments presented by Darwin in favor of a universal derivation from one primary form of all the peculiarities existing now among living beings have not made the slightest impression on my mind. "Until the facts of Nature are shown to have been mistaken by those who have collected them, and that they have a different meaning from that now generally assigned to them, I shall therefore consider the transmutation theory as a scientific mistake, untrue in its facts, unscientific in its method, and mischievous in its tendency."—Silliman's 'Journal,' July 1860, pp. 143, 154. Extract from the 3rd vol. of 'Contributions to the Natural History of the United States.'

only a very little way.* On the other hand, Lyell, up to that time a pillar of the anti-transmutationists (who regarded him, ever afterwards, as Pallas Athene may have looked at Dian, after the Endymion affair), declared himself a Darwinian, though not without putting in a serious *caveat*. Nevertheless, he was a tower of strength, and his courageous stand for truth as against consistency, did him infinite honour. As evolutionists, *sans phrase*, I do not call to mind among the biologists more than Asa Gray, who fought the battle splendidly in the United States ; Hooker, who was no less vigorous here ; the present Sir John Lubbock and my-self. Wallace was far away in the Malay Archipelago ; but, apart from his direct share in the promulgation of the theory of natural selection, no enumeration of the influences at work, at the time I am speaking of, would be complete without the mention of his powerful essay ' On the Law which has regulated the Introduction of New Species,' which was published in 1855. On reading it afresh, I have been astonished to recollect how small was the impression it made.

In France, the influence of Elie de Beaumont and of Flourens, —the former of whom is said to have " damned himself to everlasting fame" by inventing the nickname of " *la science moussante* " for Evolutionism,†—to say nothing of the ill-will of other powerful members of the Institut, produced for a

* " I see no serious objections to the formation of varieties by natural selection in the existing world, and that, so far as earlier epochs are concerned, this law may be assumed to explain the origin of closely allied species, supposing for this purpose a very long period of time.

" With regard to simple varieties and closely allied species, I believe that Mr. Darwin's theory may explain many things, and throw a great light upon numerous questions."—' Sur l'Origine de l'Espèce. Par Charles Darwin.' 'Archives des Sc. de la Bibliothèque Universelle de Genève,' pp. 242, 243, Mars 1860.

† One is reminded of the effect of another small academic epigram. The so-called vertebral theory of the skull is said to have been nipped in the bud in France by the whisper of an academician to his neighbour, that, in that case, one's head was a " *vertèbre pensante*."

long time the effect of a conspiracy of silence ; and many years passed before the Academy redeemed itself from the reproach that the name of Darwin was not to be found on the list of its members.   However, an accomplished writer, out of the range of academical influences, M. Laugel, gave an excellent and appreciative notice of the ' Origin' in the 'Revue des Deux Mondes.'   Germany took time to consider ; Bronn produced a slightly Bowdlerized translation of the ' Origin' ;. and 'Kladderadatsch' cut his jokes upon the ape origin of man ; but I do not call to mind that any scientific notability declared himself publicly in 1860.*   None of us dreamed that, in the course of a few years, the strength (and perhaps I may add the weakness) of " Darwinismus " would have its most extensive and most brilliant illustrations in the land of learning. If a foreigner may presume to speculate on the cause of this curious interval of silence, I fancy it was that one moiety of the German biologists were orthodox at any price, and the other moiety as distinctly heterodox.   The latter were evolutionists, *a priori*, already, and they must have felt the disgust natural to deductive philosophers at being offered an inductive and experimental foundation for a conviction which they had reached by a shorter cut.   It is undoubtedly trying to learn that, though your conclusions may be all right, your reasons for them are all wrong, or, at any rate, insufficient.

On the whole, then, the supporters of Mr. Darwin's views in 1860 were numerically extremely insignificant.   There is not the slightest doubt that, if a general council of the Church scientific had been held at that time, we should have been condemned by an overwhelming majority.   And there is as little doubt that, if such a council gathered now, the decree would be of an exactly contrary nature.   It would indicate a lack

---

* However, the man who stands next to Darwin in his influence on modern biologists, K. E. von Bär, wrote to me, in August 1860, expressing his general assent to evolutionist views. His phrase, " J'ai énoncé les mêmes idées . . . que M. Darwin " (vol. ii. p. 329), is. shown by his subsequent writings. to mean no more than this.

of sense, as well as of modesty, to ascribe to the men of that
generation less capacity or less honesty than their successors
possess. What, then, are the causes which led instructed and
fair-judging men of that day to arrive at a judgment so
different from that which seems just and fair to those who
follow them ? That is really one of the most interesting of all
questions connected with the history of science, and I shall
try to answer it. I am afraid that in order to do so I must
run the risk of appearing egotistical. However, if I tell my
own story it is only because I know it better than that of
other people.

I think I must have read the 'Vestiges' before I left England in 1846 ; but, if I did, the book made very little impression upon me, and I was not brought into serious contact with
the 'Species' question until after 1850. At that time, I had
long done with the Pentateuchal cosmogony, which had been
impressed upon my childish understanding as Divine truth,
with all the authority of parents and instructors, and from
which it had cost me many a struggle to get free. But my
mind was unbiassed in respect of any doctrine which presented
itself, if it professed to be based on purely philosophical and
scientific reasoning. It seemed to me then (as it does now)
that " creation," in the ordinary sense of the word, is perfectly
conceivable. I find no difficulty in imagining that, at some
former period, this universe was not in existence ; and that
it made its appearance in six days (or instantaneously, if
that is preferred), in consequence of the volition of some pre-
existent Being. Then, as now, the so-called *a priori* arguments against Theism, and, given a Deity, against the
possibility of creative acts, appeared to me to be devoid of
reasonable foundation. I had not then, and I have not now,
the smallest *a priori* objection to raise to the account of the
creation of animals and plants given in 'Paradise Lost,' in
which Milton so vividly embodies the natural sense of Genesis.
Far be it from me to say that it is untrue because it is impos-

188 ON THE RECEPTION OF

sible. I confine myself to what must be regarded as a modest
and reasonable request for some particle of evidence that the
existing species of animals and plants did originate in that
way, as a condition of my belief in a statement which appears
to me to be highly improbable.

And, by way of being perfectly fair, I had exactly the
same answer to give to the evolutionists of 1851-8. Within
the ranks of the biologists, at that time, I met with nobody,
except Dr. Grant, of University College, who had a word to say
for Evolution—and his advocacy was not calculated to advance
the cause. Outside these ranks, the only person known to me
whose knowledge and capacity compelled respect, and who
was, at the same time, a thorough-going evolutionist, was Mr.
Herbert Spencer, whose acquaintance I made, I think, in
1852, and then entered into the bonds of a friendship which,
I am happy to think, has known no interruption. Many and
prolonged were the battles we fought on this topic. But
even my friend's rare dialectic skill and copiousness of
apt illustration could not drive me from my agnostic position.
I took my stand upon two grounds : firstly, that up to that
time, the evidence in favour of transmutation was wholly
insufficient ; and, secondly, that no suggestion respecting
the causes of the transmutation assumed, which had been
made, was in any way adequate to explain the phenomena.
Looking back at the state of knowledge at that time, I
really do not see that any other conclusion was justifiable.

In those days I had never even heard of Treviranus'
'Biologie.' However, I had studied Lamarck attentively and
I had read the 'Vestiges' with due care ; but neither of them
afforded me any good ground for changing my negative
and critical attitude. As for the 'Vestiges,' I confess that
the book simply irritated me by the prodigious ignorance
and thoroughly unscientific habit of mind manifested by the
writer. If it had any influence on me at all, it set me
against Evolution ; and the only review I ever have qualms

of conscience about, on the ground of needless savagery, is one I wrote on the 'Vestiges' while under that influence.

With respect to the 'Philosophie Zoologique,' it is no reproach to Lamarck to say that the discussion of the Species question in that work, whatever might be said for it in 1809, was miserably below the level of the knowledge of half a century later. In that interval of time the elucidation of the structure of the lower animals and plants had given rise to wholly new conceptions of their relations ; histology and embryology, in the modern sense, had been created ; physiology had been reconstituted ; the facts of distribution, geological and geographical, had been prodigiously multiplied and reduced to order. To any biologist whose studies had carried him beyond mere species-mongering in 1850, one-half of Lamarck's arguments were obsolete and the other half erroneous, or defective, in virtue of omitting to deal with the various classes of evidence which had been brought to light since his time. Moreover his one suggestion as to the cause of the gradual modification of species—effort excited by change of conditions—was, on the face of it, inapplicable to the whole vegetable world. I do not think that any impartial judge who reads the 'Philosophie Zoologique' now, and who afterwards takes up Lyell's trenchant and effectual criticism (published as far back as 1830), will be disposed to allot to Lamarck a much higher place in the establishment of biological evolution than that which Bacon assigns to himself in relation to physical science generally,—*buccinator tantum.**

But, by a curious irony of fate, the same influence which led me to put as little faith in modern speculations on this subject, as in the venerable traditions recorded in the first two chapters of Genesis, was perhaps more potent than any other

* Erasmus Darwin first promulgated Lamarck's fundamental conceptions, and, with greater logical consistency, he had applied them to plants. But the advocates of his claims have failed to show that he, in any respect, anticipated the central idea of the 'Origin of Species.'

in keeping alive a sort of pious conviction that Evolution, after all, would turn out true. I have recently read afresh the first edition of the 'Principles of Geology'; and when I consider that this remarkable book had been nearly thirty years in everybody's hands, and that it brings home to any reader of ordinary intelligence a great principle and a great fact—the principle, that the past must be explained by the present, unless good cause be shown to the contrary ; and the fact, that, so far as our knowledge of the past history of life on our globe goes, no such cause can be shown *—I cannot but believe that Lyell, for others, as for myself, was the chief agent in smoothing the road for Darwin. For consistent uniformitarianism postulates evolution as much in the organic as in the inorganic world. The origin of a new species by other than ordinary agencies would be a vastly greater "catastrophe" than any of those which Lyell successfully eliminated from sober geological speculation.

In fact, no one was better aware of this than Lyell himself.† If one reads any of the earlier editions of the 'Principles' carefully (especially by the light of the interesting series of letters recently published by Sir Charles Lyell's biographer), it is easy to see that, with all his energetic opposition to Lamarck,

* The same principle and the same fact guide and result from all sound historical investigation. Grote's 'History of Greece' is a product of the same intellectual movement as Lyell's 'Principles.'

† Lyell, with perfect right, claims this position for himself. He speaks of having "advocated a law of continuity even in the organic world, so far as possible without adopting Lamarck's theory of transmutation.... "But while I taught that as often as certain forms of animals and plants disappeared, for reasons quite intelligible to us, others took their place by virtue of a causation which was beyond our comprehension ; it remained for Darwin to accumulate proof that there is no break between the incoming and the outgoing species, that they are the work of evolution, and not of special creation. . . .

"I had certainly prepared the way in this country, in six editions of my work before the 'Vestiges of Creation' appeared in 1842 [1844], for the reception of Darwin's gradual and insensible evolution of species." —'Life and Letters,' Letter to Haeckel, vol. ii. p. 436. Nov. 23, 1868.

on the one hand, and to the ideal quasi-progressionism of
Agassiz, on the other, Lyell, in his own mind, was strongly
disposed to account for the origination of all past and present
species of living things by natural causes. But he would have
liked, at the same time, to keep the name of creation for a
natural process which he imagined to be incomprehensible.

In a letter addressed to Mantell (dated March 2, 1827),
Lyell speaks of having just read Lamarck ; he expresses his
delight at Lamarck's theories, and his personal freedom from
any objections based on theological grounds. And though he
is evidently alarmed at the pithecoid origin of man involved
in Lamarck's doctrine, he observes :—

" But, after all, what changes species may really undergo !
How impossible will it be to distinguish and lay down a line,
beyond which some of the so-called extinct species have
never passed into recent ones."

Again, the following remarkable passage occurs in the postscript of a letter addressed to Sir John Herschel in 1836 :—

" In regard to the origination of new species, I am very
glad to find that you think it probable that it may be carried
on through the intervention of intermediate causes. I left this
rather to be inferred, not thinking it worth while to offend a
certain class of persons by embodying in words what would
only be a speculation." * He goes on to refer to the criticisms
which have been directed against him on the ground that, by
leaving species to be originated by miracle, he is inconsistent
with his own doctrine of uniformitarianism ; and he leaves it

* In the same sense, see the letter to Whewell, March 7, 1837, vol. ii., p. 5 :—
" In regard to this last subject [the changes from one set of animal and vegetable species to another] ... you remember what Herschel said in his letter to me. If I had stated as plainly as he has done the possibility of the introduction or origina- tion of fresh species being a natural, in contradistinction to a miraculous process, I should have raised a host of prejudices against me, which are unfortunately opposed at every step to any philosopher who attempts to address the public on these mys- terious subjects." See also letter to Sedgwick, Jan. 20, 1838, vol. ii. p. 35.

to be understood that he had not replied, on the ground of his general objection to controversy.

Lyell's contemporaries were not without some inkling of his esoteric doctrine. Whewell's 'History of the Inductive Sciences,' whatever its philosophical value, is always worth reading and always interesting, if under no other aspect than that of an evidence of the speculative limits within which a highly-placed divine might, at that time, safely range at will. In the course of his discussion of uniformitarianism, the encyclopædic Master of Trinity observes :—

" Mr. Lyell, indeed, has spoken of an hypothesis that 'the successive creation of species may constitute a regular part of the economy of nature,' but he has nowhere, I think, so described this process as to make it appear in what department of science we are to place the hypothesis. Are these new species created by the production, at long intervals, of an offspring different in species from the parents ? Or are the species so created produced without parents ? Are they gradually evolved from some embryo substance ? Or do they suddenly start from the ground, as in the creation of the poet ? . . .

" Some selection of one of these forms of the hypothesis, rather than the others, with evidence for the selection, is requisite to entitle us to place it among the known causes of change, which in this chapter we are considering. The bare conviction that a creation of species has taken place, whether once or many times, so long as it is unconnected with our organical sciences, is a tenet of Natural Theology rather than of Physical Philosophy." *

The earlier part of this criticism appears perfectly just and appropriate ; but, from the concluding paragraph, Whewell evidently imagines that by " creation " Lyell means a preternatural intervention of the Deity ; whereas the letter to Herschel shows that, in his own mind, Lyell meant natural

* Whewell's ' History,' vol. iii. p. 639–640 (ed. 2, 1847).

causation ; and I see no reason to doubt * that, if Sir Charles could have avoided the inevitable corollary of the pithecoid origin of man—for which, to the end of his life, he entertained a profound antipathy—he would have advocated the efficiency of causes now in operation to bring about the condition of the organic world, as stoutly as he championed that doctrine in reference to inorganic nature.

The fact is, that a discerning eye might have seen that some form or other of the doctrine of transmutation was inevitable, from the time when the truth enunciated by William

* The following passages in Lyell's letters appear to me decisive on this point :—

To Darwin, Oct. 3, 1859 (ii. 325), on first reading the ' Origin.'

" I have long seen most clearly that if any concession is made, all that you claim in your concluding pages will follow.

" It is this which has made me so long hesitate, always feeling that the case of Man and his Races, and of other animals, and that of plants, is one and the same, and that if a *vera causa* be admitted for one instant, [instead] of a purely unknown and imaginary one, such as the word ' creation,' all the consequences must follow."

To Darwin, March 15, 1863 (vol. ii. p. 365).

" I remember that it was the conclusion he [Lamarck] came to about man that fortified me thirty years ago against the great impression which his arguments at first made on my mind, all the greater because Constant Prévost, a pupil of Cuvier's forty years ago, told me his conviction ' that Cuvier thought species not real, but that science could not

advance without assuming that they were so.' "

To Hooker, March 9, 1863 (vol. ii. p. 361), in reference to Darwin's feeling about the 'Antiquity of Man.'

" He [Darwin] seems much disappointed that I do not go farther with him, or do not speak out more. I can only say that I have spoken out to the full extent of my present convictions, and even beyond my state of *feeling* as to man's unbroken descent from the brutes, and I find I am half converting not a few who were in arms against Darwin, and are even now against Huxley." He speaks of having had to abandon " old and long cherished ideas, which constituted the charm to me of the theoretical part of the science in my earlier days, when I believed with Pascal in the theory, as Hallam terms it, of ' the archangel ruined.' "

See the same sentiment in the letter to Darwin, March 11, 1863, p. 363 :—

" I think the old ' creation ' is almost as much required as ever, but of course it takes a new form if Lamarck's views improved by yours are adopted."

Smith, that successive strata are characterised by different kinds of fossil remains, became a firmly established law of nature. No one has set forth the speculative consequences of this generalisation better than the historian of the 'Inductive Sciences':—

"But the study of geology opens to us the spectacle of many groups of species which have, in the course of the earth's history, succeeded each other at vast intervals of time; one set of animals and plants disappearing, as it would seem, from the face of our planet, and others, which did not before exist, becoming the only occupants of the globe. And the dilemma then presents itself to us anew :—either we must accept the doctrine of the transmutation of species, and must suppose that the organized species of one geological epoch were transmuted into those of another by some long-continued agency of natural causes; or else, we must believe in many successive acts of creation and extinction of species, out of the common course of nature; acts which, therefore, we may properly call miraculous." *

Dr. Whewell decides in favour of the latter conclusion. And if any one had plied him with the four questions which he puts to Lyell in the passage already cited, all that can be said now is that he would certainly have rejected the first. But would he really have had the courage to say that a *Rhinoceros tichorhinus*, for instance, "was produced without parents;" or was "evolved from some embryo substance;" or that it suddenly started from the ground like Milton's lion "pawing to get free his hinder parts"? I permit myself to doubt whether even the Master of Trinity's well-tried courage— physical, intellectual, and moral—would have been equal to this feat. No doubt the sudden concurrence of half-a-ton of inorganic molecules into a live rhinoceros is conceivable, and therefore may be possible. But does such an event lie

---

* Whewell's 'History of the In-   vol. iii. p. 624–625. See, for the
ductive Sciences.' Ed. ii., 1847,   author's verdict, pp. 638–39.

:sufficiently within the bounds of probability to justify the belief in its occurrence on the strength of any attainable, or, indeed, imaginable, evidence?

In view of the assertion (often repeated in the early days of the opposition to Darwin) that he had added nothing to Lamarck, it is very interesting to observe that the possibility of a fifth alternative, in addition to the four he has stated, has not dawned upon Dr. Whewell's mind. The suggestion that new species may result from the selective action of external conditions upon the variations from their specific type which individuals present—and which we call "spontaneous," because we are ignorant of their causation—is as wholly unknown to the historian of scientific ideas as it was to biological specialists before 1858. But that suggestion is the central idea of the 'Origin of Species,' and contains the quintessence of Darwinism.

Thus, looking back into the past, it seems to me that my own position of critical expectancy was just and reasonable, and must have been taken up, on the same grounds, by many other persons. If Agassiz told me that the forms of life which had successively tenanted the globe were the incarnations of successive thoughts of the Deity; and that He had wiped out one set of these embodiments by an appalling geological catastrophe as soon as His ideas took a more advanced shape, I found myself not only unable to admit the accuracy of the deductions from the facts of paleontology, upon which this astounding hypothesis was founded, but I had to confess my want of any means of testing the correctness of his explanation of them. And besides that, I could by no means see what the explanation explained. Neither did it help me to be told by an eminent anatomist that species had succeeded one another in time, in virtue of "a continuously operative creational law." That seemed to me to be no more than saying that species had succeeded one another, in the form of a vote-catching resolution, with "law" to please the

man of science, and "creational" to draw the orthodox. So
I took refuge in that "*thätige Skepsis*" which Goethe has so
well defined ; and, reversing the apostolic precept to be all
things to all men, I usually defended the tenability of the
received doctrines, when I had to do with the transmuta-
tionists ; and stood up for the possibility of transmutation
among the orthodox—thereby, no doubt, increasing an already
current, but quite undeserved, reputation for needless combativeness.

I remember, in the course of my first interview with
Mr. Darwin, expressing my belief in the sharpness of the lines
of demarcation between natural groups and in the absence
of transitional forms, with all the confidence of youth and
imperfect knowledge.   I was not aware, at that time, that he
had then been many years brooding over the species-question ; and the humorous smile which accompanied his gentle
answer, that such was not altogether his view, long haunted
and puzzled me.   But it would seem that four or five years'
hard work had enabled me to understand what it meant ;
for Lyell,* writing to Sir Charles Bunbury (under date of
April 30, 1856), says :—

"When Huxley, Hooker, and Wollaston were at Darwin's
last week they (all four of them) ran a tilt against species—
further, I believe, than they are prepared to go."

I recollect nothing of this beyond the fact of meeting Mr.
Wollaston ; and except for Sir Charles' distinct assurance
as to "all four," I should have thought my *outrecuidance* was
probably a counterblast to Wollaston's conservatism.   With
regard to Hooker, he was already, like Voltaire's Habakkuk,
"*capable de tout*" in the way of advocating Evolution.

As I have already said, I imagine that most of those of my
contemporaries who thought seriously about the matter, were
very much in my own state of mind—inclined to say to
both Mosaists and Evolutionists, "a plague on both your

* 'Life and Letters,' vol. ii. p. 212.

houses!" and disposed to turn aside from an interminable and apparently fruitless discussion, to labour in the fertile fields of ascertainable fact. And I may, therefore, further suppose that the publication of the Darwin and Wallace papers in 1858, and still more that of the 'Origin' in 1859, had the effect upon them of the flash of light, which to a man who has lost himself in a dark night, suddenly reveals a road which, whether it takes him straight home or not, certainly goes his way. That which we were looking for, and could not find, was a hypothesis respecting the origin of known organic forms, which assumed the operation of no causes but such as could be proved to be actually at work. We wanted, not to pin our faith to that or any other speculation, but to get hold of clear and definite conceptions which could be brought face to face with facts and have their validity tested. The 'Origin' provided us with the working hypothesis we sought. Moreover, it did the immense service of freeing us for ever from the dilemma—refuse to accept the creation hypothesis, and what have you to propose that can be accepted by any cautious reasoner? In 1857, I had no answer ready, and I do not think that any one else had. A year later, we reproached ourselves with dulness for being perplexed by such an inquiry. My reflection, when I first made myself master of the central idea of the 'Origin,' was, "How extremely stupid not to have thought of that!" I suppose that Columbus' companions said much the same when he made the egg stand on end. The facts of variability, of the struggle for existence, of adaptation to conditions, were notorious enough ; but none of us had suspected that the road to the heart of the species problem lay through them, until Darwin and Wallace dispelled the darkness, and the beacon-fire of the 'Origin' guided the benighted.

Whether the particular shape which the doctrine of evolution, as applied to the organic world, took in Darwin's hands, would prove to be final or not, was, to me, a matter of indifference. In my earliest criticisms of the 'Origin' I ventured to point out that its logical foundation was insecure so long as experiments in selective breeding had not produced varieties which were more or less infertile ; and that insecurity remains up to the present time. But, with any and every critical doubt which my sceptical ingenuity could suggest, the Darwinian hypothesis remained incomparably more probable than the creation hypothesis. And if we had none of us been able to discern the paramount significance of some of the most patent and notorious of natural facts, until they were, so to speak, thrust under our noses, what force remained in the dilemma— creation or nothing? It was obvious that, hereafter, the probability would be immensely greater, that the links of natural causation were hidden from our purblind eyes, than that natural causation should be incompetent to produce all the phenomena of nature. The only rational course for those who had no other object than the attainment of truth, was to accept "Darwinism" as a working hypothesis, and see what could be made of it. Either it would prove its capacity to elucidate the facts of organic life, or it would break down under the strain. This was surely the dictate of common sense ; and, for once, common sense carried the day. The result has been that complete *volte-face* of the whole scientific world,. which must seem so surprising to the present generation. I do not mean to say that all the leaders of biological science have avowed themselves Darwinians ; but I do not think that there is a single zoologist, or botanist, or palæontologist, among the multitude of active workers of this generation, who is other than an evolutionist, profoundly influenced by Darwin's views. Whatever may be the ultimate fate of the particular theory put forth by Darwin, I venture to affirm that,. so far as my knowledge goes, all the ingenuity and all the learning of hostile critics has not enabled them to adduce a solitary fact, of which it can be said, this is irreconcilable with the Darwinian theory. In the prodigious variety and complexity of organic nature, there are multitudes of phenomena which are not deducible from any generalisations we have yet reached. But the same may be said of every other class of natural objects. I believe that astronomers cannot yet get the moon's motions into perfect accordance with the theory of gravitation.

It would be inappropriate, even if it were possible, to discuss the difficulties and unresolved problems which have hitherto met the evolutionist, and which will probably continue to puzzle him for many generations to come, in the course of this brief history of the reception of Mr. Darwin's great work. But there are two or three objections of a more general character, based, or supposed to be based, upon philosophical and theological foundations, which were loudly expressed in the early days of the Darwinian controversy, and which, though they have been answered over and over again, crop up now and then at the present day.

The most singular of these, perhaps immortal, fallacies, which live on, Tithonus-like, when sense and force have long deserted them, is that which charges Mr. Darwin with having attempted to reinstate the old pagan goddess, Chance. It is said that he supposes variations to come about "by chance," and that the fittest survive the "chances" of the struggle for existence, and thus "chance" is substituted for providential design.

It is not a little wonderful that such an accusation as this should be brought against a writer who has, over and over again, warned his readers that when he uses the word "spontaneous," he merely means that he is ignorant of the cause of that which is so termed; and whose whole theory crumbles to pieces if the uniformity and regularity of natural causation for illimitable past ages is denied. But probably the best answer to those who talk of Darwinism meaning the reign of "chance," is to ask them what they themselves understand by "chance." Do they believe that anything in this universe happens without reason or without a cause? Do they really conceive that any event has no cause, and could not have been predicted by any one who had a sufficient insight into the order of Nature? If they do, it is they who are the inheritors of antique superstition and ignorance, and whose minds have never been illumined by a ray of scientific thought. The one act of faith in the convert to science, is the confession of the universality of order and of the absolute validity, in all times and under all circumstances, of the law of causation. This confession is an act of faith, because, by the nature of the case, the truth of such propositions is not susceptible of proof. But such faith is not blind, but reasonable; because it is invariably confirmed by experience, and constitutes the sole trustworthy foundation for all action.

If one of these people, in whom the chance-worship of our remoter ancestors thus strangely survives, should be within reach of the sea when a heavy gale is blowing, let him betake himself to the shore and watch the scene. Let him note the infinite variety of form and size of the tossing waves out at sea; or of the curves of their foam-crested breakers, as they dash against the rocks; let him listen to the roar and scream of the shingle as it is cast up and torn down the beach; or look at the flakes of foam as they drive hither and thither before the wind; or note the play of colours, which answers a gleam of sunshine as it falls upon their myriad bubbles. Surely here, if anywhere, he will say that chance is supreme, and bend the knee as one who has entered the very penetralia of his divinity. But the man of science knows that here, as everywhere, perfect order is manifested; that there is not a curve of the waves, not a note in the howling chorus, not a rainbow-glint on a bubble, which is other than a necessary consequence of the ascertained laws of nature; and that with a sufficient knowledge of the conditions, competent physico-mathematical skill could account for, and indeed predict, every one of these "chance" events.

A second very common objection to Mr. Darwin's views was (and is), that they abolish Teleology, and eviscerate the argument from design. It is nearly twenty years since I ventured to offer some remarks on this subject, and as my arguments have as yet received no refutation, I hope I may be excused for reproducing them. I observed, "that the doctrine of Evolution is the most formidable opponent of all the commoner and coarser forms of Teleology. But perhaps the most remarkable service to the philosophy of Biology rendered by Mr. Darwin is the reconciliation of Teleology and Morphology, and the explanation of the facts of both, which his views offer. The teleology which supposes that the eye, such as we see it in man, or one of the higher vertebrata, was made with the precise structure it exhibits, for the purpose of enabling the animal which possesses it to see, has undoubtedly received its death-blow. Nevertheless, it is necessary to remember that there is a wider teleology which is not touched by the doctrine of Evolution, but is actually based upon the fundamental proposition of Evolution. This proposition is that the whole world, living and not living, is the result of the mutual interaction, according to definite laws, of the forces * possessed by the molecules of which the primitive nebulosity of the universe was composed. If this be true, it is no less certain that the existing world lay potentially in the cosmic vapour, and that a sufficient intelligence could, from a knowledge of the properties of the molecules of that vapour, have predicted, say the state of the fauna of Britain in 1869, with as much certainty as one can say what will happen to the vapour of the breath on a cold winter's day. . . . .

. . . . The teleological and the mechanical views of nature are not, necessarily, mutually exclusive. On the contrary, the more purely a mechanist the speculator is, the more firmly

* I should now like to substitute the word powers for "forces."

does he assume a primordial molecular arrangement of which all the phenomena of the universe are the consequences,. and the more completely is he thereby at the mercy of the teleologist, who can always defy him to disprove that this primordial molecular arrangement was not intended to evolve the phenomena of the universe." *

The acute champion of Teleology, Paley, saw no difficulty in admitting that the "production of things" may be the result of trains of mechanical dispositions fixed beforehand by intelligent appointment and kept in action by a power at the centre, † that is to say, he proleptically accepted the modern doctrine of Evolution; and his successors might do well to follow their leader, or at any rate to attend to his weighty reasonings, before rushing into an antagonism which has no reasonable foundation.

Having got rid of the belief in chance and the disbelief in design, as in no sense appurtenances of Evolution, the third libel upon that doctrine, that it is anti-theistic, might perhaps be left to shift for itself. But the persistence with which many people refuse to draw the plainest consequences from the propositions they profess to accept, renders it advisable to remark that the doctrine of Evolution is neither Anti-theistic nor Theistic. It simply has no more to do with Theism than the first book of Euclid has. It is quite certain that a normal fresh-laid egg contains neither cock nor hen; and it is also as certain as any proposition in physics or morals, that if such an egg is kept under proper conditions for three weeks, a cock or hen chicken will be found in it. It is also quite certain that if the shell were transparent we should be able to watch the formation of the young fowl, day by day, by a process of evolution, from a microscopic cellular germ to its full size and complication of structure. Therefore

---

* The "Genealogy of Animals"   † 'Natural Theology,' chap.. ('The Academy,' 1869), reprinted   xxiii. in ' Critiques and Addresses.'

Evolution, in the strictest sense, is actually going on in this and analogous millions and millions of instances, wherever living creatures exist. Therefore, to borrow an argument from Butler, as that which now happens must be consistent with the attributes of the Deity, if such a Being exists, Evolution must be consistent with those attributes. And, if so, the evolution of the universe, which is neither more nor less explicable than that of a chicken, must also be consistent with them. The doctrine of Evolution, therefore, does not even come into contact with Theism, considered as a philosophical doctrine. That with which it does collide, and with which it is absolutely inconsistent, is the conception of creation, which theological speculators have based upon the history narrated in the opening of the book of Genesis.

There is a great deal of talk and not a little lamentation about the so-called religious difficulties which physical science has created. In theological science, as a matter of fact, it has created none. Not a solitary problem presents itself to the philosophical Theist, at the present day, which has not existed from the time that philosophers began to think out the logical grounds and the logical consequences of Theism. All the real or imaginary perplexities which flow from the conception of the universe as a determinate mechanism, are equally involved in the assumption of an Eternal, Omnipotent and Omniscient Deity. The theological equivalent of the scientific conception of order is Providence ; and the doctrine of determinism follows as surely from the attributes of foreknowledge assumed by the theologian, as from the universality of natural causation assumed by the man of science. The angels in 'Paradise Lost' would have found the task of enlightening Adam upon the mysteries of " Fate, Foreknowledge, and Free-will," not a whit more difficult, if their pupil had been educated in a " Real-schule " and trained in every laboratory of a modern university. In respect of the great problems of Philosophy, the post-Darwinian generation is, in one sense, exactly where the præ-Darwinian generations were. They remain insoluble. But the present generation has the advantage of being better provided with the means of freeing itself from the tyranny of certain sham solutions.

The known is finite, the unknown infinite; intellectually we stand on an islet in the midst of an illimitable ocean of inexplicability. Our business in every generation is to reclaim a little more land, to add something to the extent and the solidity of our possessions. And even a cursory glance at the history of the biological sciences during the last quarter of a century is sufficient to justify the assertion, that the most potent instrument for the extension of the realm of natural knowledge which has come into men's hands, since the publication of Newton's 'Principia,' is Darwin's 'Origin of Species.'

It was badly received by the generation to which it was first addressed, and the outpouring of angry nonsense to which it gave rise is sad to think upon. But the present generation will probably behave just as badly if another Darwin should arise, and inflict upon them that which the generality of mankind most hate—the necessity of revising their convictions. Let them, then, be charitable to us ancients; and if they behave no better than the men of my day to some new benefactor, let them recollect that, after all, our wrath did not come to much, and vented itself chiefly in the bad language of sanctimonious scolds. Let them as speedily perform a strategic right-about-face, and follow the truth wherever it leads. The opponents of the new truth will discover, as those of Darwin are doing, that, after all, theories do not alter facts, and that the universe remains unaffected even though texts crumble. Or, it may be, that, as history repeats itself, their happy ingenuity will also discover that the new wine is exactly of the same vintage as the old, and that (rightly viewed) the old bottles prove to have been expressly made for holding it.

CHAPTER VI.

THE PUBLICATION OF THE 'ORIGIN OF SPECIES.'

OCTOBER 3, 1859, TO DECEMBER 31, 1859.

1859.

[UNDER the date of October 1st, 1859, in my father's Diary occurs the entry : " Finished proofs (thirteen months and ten days) of Abstract on 'Origin of Species'; 1250 copies printed. The first edition was published on November 24th, and all copies sold first day."

On October 2nd he started for a water-cure establishment at Ilkley, near Leeds, where he remained with his family until December, and on the 9th of that month he was again at Down. The only other entry in the Diary for this year is as follows : " During end of November and beginning of December, employed in correcting for second edition of 3000 copies ; multitude of letters."

The first and a few of the subsequent letters refer to proof sheets, and to early copies of the ' Origin ' which were sent to friends before the book was published.]

*C. Lyell to C. Darwin.**

October 3rd, 1859.

MY DEAR DARWIN,—I have just finished your volume and right glad I am that I did my best with Hooker to

----

* Part of this letter is given in the ' Life of Sir Charles Lyell,' vol. ii. p. 325.

persuade you to publish it without waiting for a time which probably could never have arrived, though you lived till the age of a hundred, when you had prepared all your facts on which you ground so many grand generalizations.

It is a splendid case of close reasoning, and long substantial argument throughout so many pages ; the condensation immense, too great perhaps for the uninitiated, but an effective and important preliminary statement, which will admit, even before your detailed proofs appear, of some occasional useful exemplification, such as your pigeons and cirripedes, of which you make such excellent use.

I mean that, when, as I fully expect, a new edition is soon called for, you may here and there insert an actual case to relieve the vast number of abstract propositions. So far as I am concerned, I am so well prepared to take your statements of facts for granted, that I do not think the " pièces justificatives " when published will make much difference, and I have long seen most clearly that if any concession is made, all that you claim in your concluding pages will follow. It is this which has made me so long hesitate, always feeling that the case of Man and his races, and of other animals, and that of plants is one and the same, and that if a " vera causa " be admitted for one, instead of a purely unknown and imaginary one, such as the word " Creation," all the consequences must follow.

I fear I have not time to-day, as I am just leaving this place, to indulge in a variety of comments, and to say how much I was delighted with Oceanic Islands—Rudimentary Organs—Embryology—the genealogical key to the Natural System, Geographical Distribution, and if I went on I should be copying the heads of all your chapters. But I will say a word of the Recapitulation, in case some slight alteration, or, at least, omission of a word or two be still possible in that.

In the first place, at p. 480, it cannot surely be said that the most eminent naturalists have rejected the view of the mutability of species? You do not mean to ignore G. St. Hilaire and Lamarck. As to the latter, you may say, that in regard to animals you substitute natural selection for volition to a certain considerable extent, but in his theory of the changes of plants he could not introduce volition; he may, no doubt, have laid an undue comparative stress on changes in physical conditions, and too little on those of contending organisms. He at least was for the universal mutability of species and for a genealogical link between the first and the present. The men of his school also appealed to domesticated varieties. (Do you mean *living* naturalists?) *

The first page of this most important summary gives the adversary an advantage, by putting forth so abruptly and crudely such a startling objection as the formation of "the eye," not by means analogous to man's reason, or rather by some power immeasurably superior to human reason, but by superinduced variation like those of which a cattle-breeder avails himself. Pages would be required thus to state an objection and remove it. It would be better, as you wish to persuade, to say nothing. Leave out several sentences, and in a future edition bring it out more fully. Between the throwing down of such a stumbling-block in the way of the reader, and the passage to the working ants, in p. 460, there are pages required; and these ants are a bathos to him before he has recovered from the shock of being called upon to believe the eye to have been brought to perfection, from a state of blindness or purblindness, by such variations as we witness. I think a little omission would greatly lessen the objectionableness of these sentences if you have not time to recast and amplify.

. . . . But these are small matters, mere spots on the sun. Your comparison of the letters retained in words, when

* In the published copies of the first edition, p. 480, the words are " eminent living naturalists."

no longer wanted for the sound, to rudimentary organs is excellent, as both are truly genealogical.

The want of peculiar birds in Madeira is a greater difficulty than seemed to me allowed for. I could cite passages where you show that variations are superinduced from the new circumstances of new colonists, which would require some Madeira birds, like those of the Galapagos, to be peculiar. There has been ample time in the case of Madeira and Porto Santo. . . .

You enclose your sheets in old MS., so the Post Office very properly charge them, as letters, 2*d*. extra. I wish all their fines on MS. were worth as much. I paid 4*s*. 6*d*. for such wash the other day from Paris, from a man who can prove 300 deluges in the valley of Seine.

With my hearty congratulations to you on your grand work, believe me,

Ever very affectionately yours,

CHAS. LYELL.

C. Darwin to C. Lyell.

Ilkley, Yorkshire,
October 11th [1859].

MY DEAR LYELL,—I thank you cordially for giving me so much of your valuable time in writing me the long letter of 3rd, and still longer of 4th. I wrote a line with the missing proof-sheet to Scarborough. I have adopted most thankfully all your minor corrections in the last chapter, and the greater ones as far as I could with little trouble. I damped the opening passage about the eye (in my bigger work I show the gradations in structure of the eye) by putting merely "complex organs." But you are a pretty Lord Chancellor to tell the barrister on one side how best to win the cause! The omission of "living" before eminent naturalists was a dreadful blunder.

Madeira and Bermuda Birds not peculiar.—You are right,
there is a screw out here; I thought no one would have
detected it; I blundered in omitting a discussion, which
I have written out in full. But once for all, let me say as an
excuse, that it was most difficult to decide what to omit.
Birds, which have struggled in their own homes, when settled
in a body, nearly simultaneously in a new country, would not
be subject to much modification, for their mutual relations
would not be much disturbed. But I quite agree with you,
that in time they ought to undergo some. In Bermuda and
Madeira they have, as I believe, been kept constant by the
frequent arrival, and the crossing with unaltered immigrants
of the same species from the main land. In Bermuda this
can be proved, in Madeira highly probable, as shown me by
letters from E. V. Harcourt. Moreover, there are ample ground
for believing that the crossed offspring of the new immigrants
(fresh blood as breeders would say), and old colonists of the
same species would be extra vigorous, and would be the most
likely to survive; thus the effects of such crossing in keeping
the old colonists unaltered would be much aided.

*On Galapagos productions having American type on view
of Creation.*—I cannot agree with you, that species if created
to struggle with American forms, would have to be created on
the American type. Facts point diametrically the other way.
Look at the unbroken and untilled ground in La Plata,
covered with European products, which have no near affinity
to the indigenous products. They are not American types
which conquer the aborigines. So in every island throughout
the world. Alph. De Candolle's result (though he does not
see its full importance), that thoroughly well naturalised
[plants] are in general very different from the aborigines
(belonging in large proportion of cases to non-indigenous
genera) is most important always to bear in mind. Once
for all, I am sure, you will understand that I thus write
dogmatically for brevity sake.

On the continued Creation of Monads.—This doctrine is superfluous (and groundless) on the theory of Natural Selection, which implies no *necessary* tendency to progression. A monad, if no deviation in its structure profitable to it under its *excessively simple* conditions of life occurred, might remain unaltered from long before the Silurian Age to the present day. I grant there will generally be a tendency to advance in complexity of organisation, though in beings fitted for very simple conditions it would be slight and slow. How could a complex organisation profit a monad? if it did not profit it there would be no advance. The Secondary Infusoria differ but little from the living. The parent monad form might perfectly well survive unaltered and fitted for its simple conditions, whilst the offspring of this very monad might become fitted for more complex conditions. The one primordial prototype of all living and extinct creatures may, it is possible, be now alive! Moreover, as you say, higher forms might be occasionally degraded, the snake Typhlops *seems* (?!) to have the habits of earth-worms. So that fresh creations of simple forms seem to me wholly superfluous.

"*Must you not assume a primeval creative power which does not act with uniformity, or how could man supervene?*"— I am not sure that I understand your remarks which follow the above. We must, under present knowledge, assume the creation of one or of a few forms in the same manner as philosophers assume the existence of a power of attraction without any explanation. But I entirely reject, as in my judgment quite unnecessary, any subsequent addition "of new powers and attributes and forces;" or of any "principle of improvement," except in so far as every character which is naturally selected or preserved is in some way an advantage or improvement, otherwise it would not have been selected. If I were convinced that I required such additions to the theory of natural selection, I would reject it as rubbish, but I have firm faith in it, as I cannot believe, that if false, it would explain so

many whole classes of facts, which, if I am in my senses, it
seems to explain. As far as I understand your remarks and
illustrations, you doubt the possibility of gradations of intellectual powers. Now, it seems to me, looking to existing
animals alone, that we have a very fine gradation in the intellectual powers of the Vertebrata, with one rather wide gap (not
half so wide as in many cases of corporeal structure), between
say a Hottentot and an Ourang, even if civilised as much
mentally as the dog has been from the wolf. I suppose that
you do not doubt that the intellectual powers are as important
for the welfare of each being as corporeal structure ; if so, I
can see no difficulty in the most intellectual individuals of a
species being continually selected ; and the intellect of the
new species thus improved, aided probably by effects of
inherited mental exercise. I look at this process as now
going on with the races of man ; the less intellectual races
being exterminated. But there is not space to discuss this
point. If I understand you, the turning-point in our difference
must be, that you think it impossible that the intellectual
powers of a species should be much improved by the continued natural selection of the most intellectual individuals.
To show how minds graduate, just reflect how impossible
every one has yet found it, to define the difference in mind
of man and the lower animals ; the latter seem to have the
very same attributes in a much lower stage of perfection than
the lowest savage. I would give absolutely nothing for the
theory of Natural Selection, if it requires miraculous additions
at any one stage of descent. I think Embryology, Homology, Classification, &c. &c., show us that all vertebrata have
descended from one parent ; how that parent appeared we
know not. If you admit in ever so little a degree, the
explanation which I have given of Embryology, Homology
and Classification, you will find it difficult to say : thus far
the explanation holds good, but no further ; here we must
call in "the addition of new creative forces." I think you

will be driven to reject all or admit all : I fear by your letter
it will be the former alternative ; and in that case I shall feel
sure it is my fault, and not the theory's fault, and this will
certainly comfort me. With regard to the descent of the
great Kingdoms (as Vertebrata, Articulata, &c.) from one
parent, I have said in the conclusion, that mere analogy
makes me think it probable ; my arguments and facts are
sound in my judgment only for each separate kingdom.

*The forms which are beaten inheriting some inferiority in
common.*—I dare say I have not been guarded enough, but
might not the term inferiority include less perfect adaptation
to physical conditions ?

My remarks apply not to single species, but to groups or
genera ; the species of most genera are adapted at least to
rather hotter, and rather less hot, to rather damper and dryer
climates ; and when the several species of a group are beaten
and exterminated by the several species of another group, it
will not, I think, generally be from *each* new species being
adapted to the climate, but from all the new species having
some common advantage in obtaining sustenance, or escaping
enemies. As groups are concerned, a fairer illustration than
negro and white in Liberia would be the almost certain future
extinction of the genus ourang by the genus man, not owing
to man being better fitted for the climate, but owing to the
inherited intellectual inferiority of the Ourang-genus to Man-
genus, by his intellect, inventing fire-arms and cutting down
forests. I believe, from reasons given in my discussion, that
acclimatisation is readily effected under nature. It has taken
me so many years to disabuse my mind of the *too* great importance of climate—its important influence being so conspicuous,
whilst that of a struggle between creature and creature is so
hidden—that I am inclined to swear at the North Pole, and
as Sydney Smith said, even to speak disrespectfully of the
Equator. I beg you often to reflect (I have found *nothing*
so instructive) on the case of thousands of plants in the

middle point of their respective ranges, and which, as we positively know, can perfectly well withstand a little more heat and cold, a little more damp and dry, but which in the metropolis of their range do not exist in vast numbers, although, if many of the other inhabitants were destroyed [they] would cover the ground. We thus clearly see that their numbers are kept down, in almost every case, not by climate, but by the struggle with other organisms. All this you will perhaps think very obvious ; but, until I repeated it to myself thousands of times, I took, as I believe, a wholly wrong view of the whole economy of nature. . . .

Hybridism.—I am so much pleased that you approve of this chapter ; you would be astonished at the labour this cost me ; so often was I, on what I believe was, the wrong scent.

Rudimentary Organs.—On the theory of Natural Selection there is a wide distinction between Rudimentary Organs and what you call germs of organs, and what I call in my bigger book "nascent" organs. An organ should not be called rudimentary unless it be useless—as teeth which never cut through the gums—the papillæ, representing the pistil in male flowers, wing of Apteryx, or better, the little wings under soldered elytra. These organs are now plainly useless, and *à fortiori*, they would be useless in a less developed state. Natural Selection acts exclusively by preserving successive slight, *useful* modifications. Hence Natural Selection cannot possibly make a useless or rudimentary organ. Such organs are solely due to inheritance (as explained in my discussion), and plainly bespeak an ancestor having the organ in a useful condition. They may be, and often have been, worked in for other purposes, and then they are only rudimentary for the original function, which is sometimes plainly apparent. A nascent organ, though little developed, as it has to be developed must be useful in every stage of development. As we cannot prophesy, we cannot tell what organs are now nascent ; and nascent organs will rarely have been handed down by certain members of a class from a remote period to the present day, for beings with any important organ but little developed, will generally have been supplanted by their descendants with the organ well developed.   The mammary glands in Ornithorhynchus may, perhaps, be considered as nascent compared with the udders of a cow—*Ovigerous frena*, in certain cirripedes, are nascent branchiæ—in [illegible] the swim bladder is almost rudimentary for this purpose, and is nascent as a lung.   The small wing of penguin, used only as a fin, might be nascent as a wing; not that I think so; for the whole structure of the bird is adapted for flight, and a penguin so closely resembles other birds, that we may infer that its wings have probably been modified, and reduced by natural selection, in accordance with its sub-aquatic habits.   Analogy thus often serves as a guide in distinguishing whether an organ is rudimentary or nascent.   I believe the *Os coccyx* gives attachment to certain muscles, but I cannot doubt that it is a rudimentary tail. The bastard wing of birds is a rudimentary digit; and I believe that if fossil birds are found very low down in the series, they will be seen to have a double or bifurcated wing. Here is a bold prophecy!

To admit prophetic germs, is tantamount to rejecting the theory of Natural Selection.

I am very glad you think it worth while to run through my book again, as much, or more, for the subject's sake as for my own sake.   But I look at your keeping the subject for some little time before your mind—raising your own difficulties and solving them—as far more important than reading my book.   If you think enough, I expect you will be perverted, and if you ever are, I shall know that the theory of Natural Selection is, in the main, safe; that it includes, as now put forth, many errors, is almost certain, though I cannot see them. Do not, of course, think of answering this; but if you have other *occasion* to write again, just say whether I have, in ever

so slight a degree, shaken any of your objections.   Farewell.
With my cordial thanks for your long letters and valuable
remarks,

Believe me, yours most truly,
C. DARWIN.

P.S.—You often allude to Lamarck's work; I do not know
what you think about it, but it appeared to me extremely
poor; I got not a fact or idea from it.

C. Darwin to L. Agassiz.*

Down, November 11th [1859].

MY DEAR SIR,—I have ventured to send you a copy of my
book (as yet only an abstract) on the 'Origin of Species.'
As the conclusions at which I have arrived on several points
differ so widely from yours, I have thought (should you at
any time read my volume) that you might think that I had
sent it to you out of a spirit of defiance or bravado; but I
assure you that I act under a wholly different frame of mind.
I hope that you will at least give me credit, however erroneous you may think my conclusions, for having earnestly
endeavoured to arrive at the truth.   With sincere respect,
I beg leave to remain,

Yours very faithfully,
CHARLES DARWIN.

---

* Jean Louis Rodolphe Agassiz, born at Mortier, on the lake of Morat in Switzerland, on May 28, 1807. He emigrated to America in 1846, where he spent the rest of his life, and died Dec. 14, 1873.   His 'Life,' written by his widow, was published in 1885.   The following extract from a letter to Agassiz (1850) is worth giving, as showing how my father regarded him, and it may be added that his cordial feelings towards the great American naturalist remained strong to the end of his life :—

" I have seldom been more deeply gratified than by receiving your most kind present of 'Lake Su-perior.'   I had heard of it, and had much wished to read it, but I con-fess that it was the very great honour of having in my posses-sion a work with your autograph as a presentation copy, that has given me such lively and sincere pleasure.   I cordially thank you for it.   I have begun to read it with uncommon interest, which I see will increase as I go on."

C. Darwin to A. De Candolle.

Down, November 11th [1859].

DEAR SIR,—I have thought that you would permit me to
send you (by Messrs. Williams and Norgate, booksellers)
a copy of my work (as yet only an abstract) on the 'Origin
of Species.' I wish to do this, as the only, though quite
inadequate manner, by which I can testify to you the extreme
interest which I have felt, and the great advantage which I
have derived, from studying your grand and noble work on
Geographical Distribution. Should you be induced to read
my volume, I venture to remark that it will be intelligible
only by reading the whole straight through, as it is very much
condensed. It would be a high gratification to me if any
portion interested you. But I am perfectly well aware that
you will entirely disagree with the conclusion at which I have
arrived.

You will probably have quite forgotten me ; but many
years ago you did me the honour of dining at my house in
London to meet M. and Madame Sismondi,* the uncle and
aunt of my wife. With sincere respect, I beg to remain,

Yours very faithfully,

CHARLES DARWIN.

C. Darwin to Hugh Falconer.

Down, November 11th [1859].

MY DEAR FALCONER,—I have told Murray to send you
a copy of my book on the 'Origin of Species,' which as yet
is only an abstract.

If you read it, you must read it straight through, otherwise
from its extremely condensed state it will be unintelligible.

Lord, how savage you will be, if you read it, and how you
will long to crucify me alive ! I fear it will produce no other

---

* Jessie Allen, sister of Mrs. Josiah Wedgwood of Maer.

effect on you; but if it should stagger you in ever so slight a degree, in this case, I am fully convinced that you will become, year after year, less fixed in your belief in the immutability of species.   With this audacious and presumptuous conviction,

I remain, my dear Falconer,
Yours most truly,
CHARLES DARWIN.

C. Darwin to Asa Gray.

Down, November 11th [1859].

MY DEAR GRAY,—I have directed a copy of my book (as yet only an abstract) on the 'Origin of Species' to be sent you.   I know how you are pressed for time; but if you can read it, I shall be infinitely gratified . . . . If ever you do read it, and can screw out time to send me (as I value your opinion so highly), however short a note, telling me what you think its weakest and best parts, I should be extremely grateful.   As you are not a geologist, you will excuse my conceit in telling you that Lyell highly approves of the two Geological chapters, and thinks that on the Imperfection of the Geological Record not exaggerated.   He is nearly a convert to my views. . . . .

Let me add I fully admit that there are very many difficulties not satisfactorily explained by my theory of descent with modification, but I cannot possibly believe that a false theory would explain so many classes of facts as I think it certainly does explain.   On these grounds I drop my anchor, and believe that the difficulties will slowly disappear. . . .

C. Darwin to J. S. Henslow.

Down, November 11th, 1859.

MY DEAR HENSLOW,—I have told Murray to send a copy of my book on Species to you, my dear old master in Natural

History; I fear, however, that you will not approve of your pupil in this case. The book in its present state does not show the amount of labour which I have bestowed on the subject.

If you have time to read it carefully, and would take the trouble to point out what parts seem weakest to you and what best, it would be a most material aid to me in writing my bigger book, which I hope to commence in a few months. You know also how highly I value your judgment. But I am not so unreasonable as to wish or expect you to write detailed and lengthy criticisms, but merely a few general remarks, pointing out the weakest parts.

If you are *in even so slight a degree* staggered (which I hardly expect) on the immutability of species, then I am convinced with further reflection you will become more and more staggered, for this has been the process through which my mind has gone. My dear Henslow,

Yours affectionately and gratefully,

C. DARWIN.

C. Darwin to John Lubbock.

Ilkley, Yorkshire,
Saturday [November 12th, 1859].

. . . Thank you much for asking me to Brighton. I hope much that you will enjoy your holiday. I have told Murray to send a copy for you to Mansion House Street, and I am surprised that you have not received it. There are so many valid and weighty arguments against my notions, that you, or any one, if you wish on the other side, will easily persuade yourself that I am wholly in error, and no doubt I am in part in error, perhaps wholly so, though I cannot see the blindness of my ways. I dare say when thunder and lightning were first proved to be due to secondary causes, some regretted to

---

* The present Sir John Lubbock.

give up the idea that each flash was caused by the direct hand of God.

Farewell, I am feeling very unwell to-day, so no more.

Yours very truly,

C. DARWIN.

C. Darwin to John Lubbock.

Ilkley, Yorkshire,

Tuesday [November 15th, 1859].

MY DEAR LUBBOCK,—I beg pardon for troubling you again. I do not know how I blundered in expressing myself in making you believe that we accepted your kind invitation to Brighton. I meant merely to thank you sincerely for wishing to see such a worn-out old dog as myself. I hardly know when we leave this place,—not under a fortnight, and then we shall wish to rest under our own roof-tree.

I do not think I hardly ever admired a book more than Paley's 'Natural Theology.' I could almost formerly have said it by heart.

I am glad you have got my book, but I fear that you value it far too highly. I should be grateful for any criticisms. I care not for Reviews; but for the opinion of men like you and Hooker and Huxley and Lyell, &c.

Farewell, with our joint thanks to Mrs. Lubbock and yourself. Adios.

C. DARWIN.

C. Darwin to L. Jenyns.*

Ilkley, Yorkshire,

November 13th, 1859.

MY DEAR JENYNS,—I must thank you for your very kind note forwarded to me from Down. I have been much out of health this summer, and have been hydropathising here for the last six weeks with very little good as yet. I shall stay

---

* Now Rev. L. Blomefield.

here for another fortnight at least. Please remember that my book is only an abstract, and very much condensed, and, to be at all intelligible, must be carefully read. I shall be very grateful for any criticisms. But I know perfectly well that you will not at all agree with the lengths which I go. It took long years to convert me. I may, of course, be egregiously wrong; but I cannot persuade myself that a theory which explains (as I think it certainly does) several large classes of facts, can be wholly wrong; notwithstanding the several difficulties which have to be surmounted somehow, and which stagger me even to this day.

I wish that my health had allowed me to publish in extenso; if ever I get strong enough I will do so, as the greater part is written out, and of which MS. the present volume is an abstract.

I fear this note will be almost illegible; but I am poorly, and can hardly sit up. Farewell; with thanks for your kind note, and pleasant remembrances of good old days.

Yours very sincerely,
C. DARWIN.

C. Darwin to A. R. Wallace.

Ilkley, November 13th, 1859.

MY DEAR SIR,—I have told Murray to send you by post (if possible) a copy of my book, and I hope that you will receive it at nearly the same time with this note. (N.B. I have got a bad finger, which makes me write extra badly.) If you are so inclined, I should very much like to hear your general impression of the book, as you have thought so profoundly on the subject, and in so nearly the same channel with myself. I hope there will be some little new to you, but I fear not much. Remember it is only an abstract, and very much condensed. God knows what the public will think. No one has read it, except Lyell, with whom I have had much correspondence. Hooker thinks him a complete convert, but he does not seem so in his letters to me; but is evidently deeply interested in the subject. I do not think your share in the theory will be overlooked by the real judges, as Hooker, Lyell, Asa Gray, &c. . I have heard from Mr. Sclater that your paper on the Malay Archipelago has been read at the Linnean Society, and that he was *extremely* much interested by it.

I have not seen one naturalist for six or nine months, owing to the state of my health, and therefore I really have no news to tell you. I am writing this at Ilkley Wells, where I have been with my family for the last six weeks, and shall stay for some few weeks longer. As yet I have profited very little. God knows when I shall have strength for my bigger book.

I sincerely hope that you keep your health; I suppose that you will be thinking of returning* soon with your magnificent collections, and still grander mental materials. You will be puzzled how to publish. The Royal Society fund will be worth your consideration. With every good wish, pray believe me,

Yours very sincerely,

CHARLES DARWIN.

P.S.—I think that I told you before that Hooker is a complete convert. If I can convert Huxley I shall be content.

C. Darwin to W. D. Fox.

Ilkley, Yorkshire,
Wednesday [November 16th, 1859].

. . . . . . I like the place very much, and the children have enjoyed it much, and it has done my wife good. It did H. good at first, but she has gone back again. I have had a series of calamities; first a sprained ankle, and then a badly

* Mr. Wallace was in the Malay Archipelago.

swollen whole leg and face, much rash, and a frightful succession of boils—four or five at once. I have felt quite ill, and
have little faith in this "unique crisis," as the doctor calls it,
doing me much good. . . . . . You will probably have
received, or will very soon receive, my weariful book on
species. I naturally believe it mainly includes the truth, but
you will not at all agree with me. Dr. Hooker, whom I consider one of the best judges in Europe, is a complete convert,
and he thinks Lyell is likewise; certainly, judging from Lyell's
letters to me on the subject, he is deeply staggered. Farewell.
If the spirit moves you, let me have a line. . . .

C. Darwin to W. B. Carpenter.

Ilkley, Yorkshire,
November 18th [1859].

MY DEAR CARPENTER,—I must thank you for your letter
on my own account, and, if I know myself, still more warmly
for the subject's sake. As you seem to have understood my
last chapter without reading the previous chapters, you must
have maturely and most profoundly self-thought out the subject; for I have found the most extraordinary difficulty in
making even able men understand at what I was driving.
There will be strong opposition to my views. If I am in the
main right (of course including partial errors unseen by me),
the admission of my views will depend far more on men, like
yourself, with well-established reputations, than on my own
writings. Therefore, on the supposition that when you have
read my volume you think the view in the main true, I thank
and honour you for being willing to run the chance of unpopu-
larity by advocating the view. I know not in the least
whether any one will review me in any of the Reviews. I do
not see how an author could enquire or interfere; but if you
are willing to review me anywhere, I am sure from the admiration which I have long felt and expressed for your ' Compara-

tive Physiology,' that your review will be excellently done, and will do good service in the cause for which I think I am not selfishly deeply interested.   I am feeling very unwell to-day, and this note is badly, perhaps hardly intelligibly, expressed ; but you must excuse me, for I could not let a post pass, without thanking you for your note.   You will have a tough job even to shake in the slightest degree Sir H. Holland.   I do not think (privately I say it) that the great man has knowledge enough to enter on the subject.   Pray believe me with sincerity,

Yours truly obliged,
C. DARWIN.

P.S.—As you are not a practical geologist, let me add that Lyell thinks the chapter on the Imperfection of the Geological Record *not* exaggerated.

C. Darwin to W. B. Carpenter.

Ilkley, Yorkshire,
November 19th [1859].

MY DEAR CARPENTER,—I beg pardon for troubling you again.   If, after reading my book, you are able to come to a conclusion in any degree definite, will you think me very unreasonable in asking you to let me hear from you.   I do not ask for a long discussion, but merely for a brief idea of your general impression.   From your widely extended knowledge, habit of investigating the truth, and abilities, I should value your opinion in the very highest rank.   Though I, of course, believe in the truth of my own doctrine, I suspect that no belief is vivid until shared by others.   As yet I know only one believer, but I look at him as of the greatest authority, viz. Hooker.   When I think of the many cases of men who have studied one subject for years, and have persuaded themselves of the truth of the foolishest doctrines, I feel sometimes a little frightened, whether I may not be one of these monomaniacs.

Again pray excuse this, I fear, unreasonable request. A short note would suffice, and I could bear a hostile verdict, and shall have to bear many a one.

Yours very sincerely,

C. DARWIN.

C. Darwin to J. D. Hooker.

Ilkley, Yorkshire,

Sunday [November, 1859].

MY DEAR HOOKER,—I have just read a review on my book in the *Athenæum*,* and it excites my curiosity much who is the author. If you should hear who writes in the *Athenæum* I wish you would tell me. It seems to me well done, but the reviewer gives no new objections, and, being hostile, passes over every single argument in favour of the doctrine, . . . I fear from the tone of the review, that I have written in a conceited and cocksure style,† which shames me a little. There is another review of which I should like to know the author, viz. of H. C. Watson in the *Gardeners' Chronicle*.‡ Some of the remarks are like yours, and he does deserve punishment; but surely the review is too severe. Don't you think so ? . . . .

I have heard from Carpenter, who, I think, is likely to be a convert. Also from Quatrefages, who is inclined to go a long way with us. He says that he exhibited in his lecture a diagram closely like mine !

---

* Nov. 19, 1859.

† The Reviewer speaks of the author's " evident self-satisfaction," and of his disposing of all diffic- ulties "more or less confidently."

‡ A review of the fourth volume of Watson's 'Cybele Britannica,' *Gard. Chron.*, 1859, p. 911.

I shall stay here one fortnight more, and then go to Down, staying on the road at Shrewsbury a week.   I have been very unfortunate : out of seven weeks I have been confined for five to the house.   This has been bad for me, as I have not been able to help thinking to a foolish extent about my book.   If some four or five *good* men came round nearly to our view, I shall not fear ultimate success.   I long to learn what Huxley thinks.   Is your Introduction* published ?   I suppose that you will sell it separately.   Please answer this, for I want an extra copy to send away to Wallace.   I am very bothersome, farewell.

Yours affectionately,
C. DARWIN.

I was very glad to see the Royal Medal for Mr. Bentham.

C. Darwin to J. D. Hooker.

Down [November 21st, 1859].

MY DEAR HOOKER,—Pray give my thanks to Mrs. Hooker for her extremely kind note, which has pleased me much. We are very sorry she cannot come here, but shall be delighted to see you and W. (our boys will be at home) here in the 2nd week of January, or any other time.   I shall much enjoy discussing any points in my book with you. . . .

I hate to hear you abuse your own work.   I, on the contrary, so sincerely value all that you have written.   It is an old and firm conviction of mine, that the Naturalists who accumulate facts and make many partial generalisations are the *real* benefactors of science.   Those who merely accumulate facts I cannot very much respect.

I had hoped to have come up for the Club to-morrow, but very much doubt whether I shall be able.   Ilkley seems to have done me no essential good.   I attended the Bench on

* Introduction to the ' Flora of Australia.'

VOL. II.   Q

Monday, and was detained in adjudicating some troublesome
cases 1½ hours longer than usual, and came home utterly
knocked up, and cannot rally.   I am not worth an old
button. . . . . Many thanks for your pleasant note.

Ever yours,

C. DARWIN.

P.S.—I feel confident that for the future progress of the
subject of the origin, and manner of formation of species, the
assent and arguments and facts of working naturalists, like
yourself, are far more important than my own book ; so for
God's sake do not abuse your Introduction.

H. C. Watson to C. Darwin.

Thames Ditton, November 21st [1859].

MY DEAR SIR,—Once commenced to read the 'Origin,' I
could not rest till I had galloped through the whole.   I shall
now begin to re-read it more deliberately.   Meantime I am
tempted to write you the first impressions, not doubting that
they will, in the main, be the permanent impressions :—

1st. Your leading idea will assuredly become recognised as
an established truth in science, *i.e.* "Natural selection."   It
has the characteristics of all great natural truths, clarifying
what was obscure, simplifying what was intricate, adding
greatly to previous knowledge.   You are the greatest revo-
lutionist in natural history of this century, if not of all
centuries.

2nd. You will perhaps need, in some degree, to limit or
modify, possibly in some degree also to extend, your present
applications of the principle of natural selection.   Without
going to matters of more detail, it strikes me that there is
one considerable primary inconsistency, by one failure in the
analogy between varieties and species ; another by a sort of
barrier assumed for nature on insufficient grounds, and arising
from "divergence."   These may, however, be faults in my

own mind, attributable to yet incomplete perception of your views. And I had better not trouble you about them before again reading the volume.

3rd. Now these novel views are brought fairly before the scientific public, it seems truly remarkable how so many of them could have failed to see their right road sooner. How could Sir C. Lyell, for instance, for thirty years read, write, and think, on the subject of species *and their succession*, and yet constantly look down the wrong road!

A quarter of a century ago, you and I must have been in something like the same state of mind on the main question. But you were able to see and work out the *quo modo* of the succession, the all-important thing, while I failed to grasp it. I send by this post a little controversial pamphlet of old date—Combe and Scott. If you will take the trouble to glance at the passages scored on the margin, you will see that, a quarter of a century ago, I was also one of the few who then doubted the absolute distinctness of species, and special creations of them. Yet I, like the rest, failed to detect the *quo modo* which was reserved for your penetration to *discover*, and your discernment to *apply*.

You answered my query about the hiatus between Satyrus and Homo as was expected. The obvious explanation really never occurred to me till some months after I had read the papers in the 'Linnean Proceedings.' The first species of *Fere-homo* * would soon make direct and exterminating war upon his *Infra-homo* cousins. The gap would thus be made, and then go on increasing, into the present enormous and still widening hiatus. But how greatly this, with your chronology of animal life, will shock the ideas of many men!

Very sincerely,
HEWETT C. WATSON.

* " Almost-man."

J. D. Hooker to C. Darwin.

Athenæum, Monday [Nov. 21, 1859].

MY DEAR DARWIN,—I am a sinner not to have written you ere this, if only to thank you for your glorious book— what a mass of close reasoning on curious facts and fresh phenomena—it is capitally written, and will be very suc- cessful. I say this on the strength of two or three plunges into as many chapters, for I have not yet attempted to read it. Lyell, with whom we are staying, is perfectly enchanted, and is absolutely gloating over it. I must accept your com- pliment to me, and acknowledgment of supposed assistance from me, as the warm tribute of affection from an honest (though deluded) man, and furthermore accept it as very pleasing to my vanity ; but, my dear fellow, neither my name nor my judgment nor my assistance deserved any such com- pliments, and if I am dishonest enough to be pleased with what I don't deserve, it must just pass. How different the *book* reads from the MS. I see I shall have much to talk over with you. Those lazy printers have not finished my luckless Essay ; which, beside your book, will look like a ragged handkerchief beside a Royal Standard . . .

All well, ever yours affectionately,

JOS. D. HOOKER.

C. Darwin to J. D. Hooker.

Ilkley, Yorkshire [November, 1859].

MY DEAR HOOKER,—I cannot help it, I must thank you for your affectionate and most kind note. My head will be turned. By Jove, I must try and get a bit modest. I was a little chagrined by the review.* I hope it was *not* ——.

---

* This refers to the review in the *Athenæum*, Nov. 19, 1859, where the reviewer, after touching on the theological aspects of the book, leaves the author to "the mercies of the Divinity Hall, the College, the Lecture Room, and the Museum."

As advocate, he might think himself justified in giving the argument only on one side.   But the manner in which he drags in immortality, and sets the priests at me, and leaves me to their mercies, is base.   He would, on no account, burn me, but he will get the wood ready, and tell the black beasts how to catch me. . . . It would be unspeakably grand if Huxley were to lecture on the subject, but I can see this is a mere chance ; Faraday might think it too unorthodox.

. . . I had a letter from [Huxley] with such tremendous praise of my book, that modesty (as I am trying to cultivate that difficult herb) prevents me sending it to you, which I should have liked to have done, as he is very modest about himself.

You have cockered me up to that extent, that I now feel I can face a score of savage reviewers.   I suppose you are still with the Lyells.   Give my kindest remembrance to them.   I triumph to hear that he continues to approve.

Believe me, your would-be modest friend,

C. D.

C. Darwin to C. Lyell.

Ilkley Wells, Yorkshire,
November 23rd [1859].

MY DEAR LYELL,—You seemed to have worked admirably on the species question ; there could not have been a better plan than reading up on the opposite side.   I rejoice profoundly that you intend admitting the doctrine of modification in your new edition ;* nothing, I am convinced, could be more important for its success.   I honour you most sincerely. To have maintained in the position of a master, one side of a question for thirty years, and then deliberately give it up, is a

---

* It appears from Sir Charles Lyell's published letters that he intended to admit the doctrine of evolution in a new edition of the 'Manual,' but this was not pub-lished till 1865.   He was, however, at work on the 'Antiquity of Man' in 1860, and had already determined to discuss the 'Origin' at the end of the book.

fact to which I much doubt whether the records of science offer a parallel. For myself, also, I rejoice profoundly ; for, thinking of so many cases of men pursuing an illusion for years, often and often a cold shudder has run through me, and I have asked myself whether I may not have devoted my life to a phantasy. Now I look at it as morally impossible that investigators of truth, like you and Hooker, can be wholly wrong, and therefore I rest in peace. Thank you for criticisms, which, if there be a second edition, I will attend to. I have been thinking that if I am much execrated as an atheist, &c., whether the admission of the doctrine of natural selection could injure your works ; but I hope and think not, for, as far as I can remember, the virulence of bigotry is expended on the first offender, and those who adopt his views are only pitied as deluded, by the wise and cheerful bigots.

I cannot help thinking that you overrate the importance of the multiple origin of dogs. The only difference is, that in the case of single origins, all difference of the races has originated since man domesticated the species. In the case of multiple origins, part of the difference was produced under natural conditions. I should *infinitely* prefer the theory of single origin in all cases, if facts would permit its reception. But there seems to me some *à priori* improbability (seeing how fond savages are of taming animals), that throughout all times, and throughout all the world, man should have domesticated one single species alone, of the widely distributed genus Canis. Besides this, the close resemblance of at least three kinds of American domestic dogs to wild species still inhabiting the countries where they are now domesticated, seems to almost compel admission that more than one wild Canis has been domesticated by man.

I thank you cordially for all the generous zeal and interest you have shown about my book, and I remain, my dear Lyell,
Your affectionate friend and disciple,
CHARLES DARWIN.

Sir J. Herschel, to whom I sent a copy, is going to read my book. He says he leans to the side opposed to me. If you should meet him after he has read me, pray find out what he thinks, for, of course, he will not write; and I should excessively like to hear whether I produce any effect on such a mind.

T. H. Huxley to C. Darwin.

Jermyn Street, W.,
November 23rd, 1859.

MY DEAR DARWIN,—I finished your book yesterday, a lucky examination having furnished me with a few hours of continuous leisure.

Since I read Von Bär's * essays, nine years ago, no work on Natural History Science I have met with has made so great an impression upon me, and I do most heartily thank you for the great store of new views you have given me. Nothing, I think, can be better than the tone of the book, it impresses those who know nothing about the subject. As for your doctrine, I am prepared to go to the stake, if requisite, in support of Chapter IX., and most parts of Chapters X., XI., XII., and Chapter XIII. contains much that is most admirable, but on one or two points I enter a *caveat* until I can see further into all sides of the question.

As to the first four chapters, I agree thoroughly and fully with all the principles laid down in them. I think you have demonstrated a true cause for the production of species, and have thrown the *onus probandi*, that species did not arise in the way you suppose, on your adversaries.

But I feel that I have not yet by any means fully realized the bearings of those most remarkable and original

---

* Karl Ernst von Baer, b. 1792, d. at Dorpat 1876—one of the most distinguished biologists of the century. He practically founded the modern science of embryology.

Chapters III., IV. and V., and I will write no more about them just now.

The only objections that have occurred to me are, 1st that you have loaded yourself with an unnecessary difficulty in adopting *Natura non facit saltum* so unreservedly. . . . And 2nd, it is not clear to me why, if continual physical conditions are of so little moment as you suppose, variation should occur at all.

However, I must read the book two or three times more before I presume to begin picking holes.

I trust you will not allow yourself to be in any way disgusted or annoyed by the considerable abuse and misrepresentation which, unless I greatly mistake, is in store for you. Depend upon it you have earned the lasting gratitude of all thoughtful men. And as to the curs which will bark and yelp, you must recollect that some of your friends, at any rate, are endowed with an amount of combativeness which (though you have often and justly rebuked it) may stand you in good stead.

I am sharpening up my claws and beak in readiness.

Looking back over my letter, it really expresses so feebly all I think about you and your noble book that I am half ashamed of it ; but you will understand that, like the parrot in the story, " I think the more."

Ever yours faithfully,

T. H. HUXLEY.

C. Darwin to T. H. Huxley.

Ilkley, Nov. 25 [1859].

MY DEAR HUXLEY,—Your letter has been forwarded to me from Down. Like a good Catholic who has received extreme unction, I can now sing " nunc dimittis." I should have been more than contented with one quarter of what you have said. Exactly fifteen months ago, when I put pen to

paper for this volume, I had awful misgivings; and thought
perhaps I had deluded myself, like so many have done,
and I then fixed in my mind three judges, on whose decision
I determined mentally to abide.   The judges were Lyell,
Hooker, and yourself.   It was this which made me so excessively anxious for your verdict.   I am now contented, and
can sing my "nunc dimittis."   What a joke it would be if I
pat you on the back when you attack some immovable creationists !   You have most cleverly hit on one point, which has
greatly troubled me ; if, as I must think, external conditions
produce little *direct* effect, what the devil determines each
particular variation?   What makes a tuft of feathers come
on a cock's head, or moss on a moss-rose?   I shall much like
to talk over this with you. . . .

My dear Huxley, I thank you cordially for your letter.

Yours very sincerely,

C. DARWIN.

P.S.—Hereafter I shall be particularly curious to hear what
you think of my explanation of Embryological similarity.
On classification I fear we shall split.   Did you perceive the
argumentum ad hominem Huxley about the kangaroo and
bear ?

Erasmus Darwin to C. Darwin.

November 23rd [1859].

DEAR CHARLES,—I am so much weaker in the head, that
I hardly know if I can write, but at all events I will jot down
a few things that the Dr.* has said.   He has not read much
above half, so as he says he can give no definite conclusion,
and it is my private belief he wishes to remain in that state.
. . . He is evidently in a dreadful state of indecision, and
keeps stating that he is not tied down to either view, and that
he has always left an escape by the way he has spoken of

* Dr., afterwards Sir Henry Holland.

varieties.   I happened to speak of the eye before he had read that part, and it took away his breath—utterly impossible—structure—function, &c., &c., &c., but when he had read it he hummed and hawed, and perhaps it was partly conceivable, and then he fell back on the bones of the ear, which were beyond all probability or conceivability.   He mentioned a slight blot, which I also observed, that in speaking of the slave-ants carrying one another, you change the species without giving notice first, and it makes one turn back. . . .

. . . For myself I really think it is the most interesting book I ever read, and can only compare it to the first know-ledge of chemistry, getting into a new world or rather behind the scenes.   To me the geographical distribution, I mean the relation of islands to continents is the most convincing of the proofs, and the relation of the oldest forms to the existing species.   I dare say I don't feel enough the absence of varieties, but then I don't in the least know if everything now living were fossilized whether the palæontologists could distinguish them.   In fact the *à priori* reasoning is so entirely satisfactory to me that if the facts won't fit in, why so much the worse for the facts is my feeling.   My ague has left me in such a state of torpidity that I wish I had gone through the process of natural selection.

Yours affectionately,
E. A. D.

C. Darwin to C. Lyell.

Ilkley, November [24th, 1859].

My dear Lyell,—Again I have to thank you for a most valuable lot of criticisms in a letter dated 22nd.

This morning I heard also from Murray that he sold the whole edition * the first day to the trade. He wants a new edition instantly, and this utterly confounds me. Now, under

---

* First edition, 1250 copies.

water-cure, with all nervous power directed to the skin, I cannot possibly do head-work, and I must make only actually necessary corrections. But I will, as far as I can without my manuscript, take advantage of your suggestions : I must not attempt much. Will you send me one line to say whether I must strike out about the secondary whale,* it goes to my heart. About the rattle-snake, look to my Journal, under Trigonocephalus, and you will see the probable origin of the rattle, and generally in transitions it is the *premier pas qui coûte.*

Madame Belloc wants to translate my book into French ; I have offered to look over proofs for *scientific* errors. Did you ever hear of her ? I believe Murray has agreed at my urgent advice, but I fear I have been rash and premature. Quatrefages has written to me, saying he agrees largely with my views. He is an excellent naturalist. I am pressed for time. Will you give us one line about the whales? Again I thank you for never-tiring advice and assistance ; I do in truth reverence your unselfish and pure love of truth.

My dear Lyell, ever yours,

C. DARWIN.

[With regard to a French translation, he wrote to Mr. Murray in Nov. 1859: "I am *extremely* anxious, for the subject's sake (and God knows not for mere fame), to have my book translated ; and indirectly its being known abroad will do good to the English sale. If it depended on me, I should agree without payment, and instantly send a copy, and only beg that she [Mme. Belloc] would get some scientific man to look over the translation. . . . You might say that, though I am a very poor French scholar, I could detect any scientific mistake, and would read over the French proofs."

The proposed translation was not made, and a second plan fell through in the following year. He wrote to M. de

---

* The passage was omitted in the second edition.

Quatrefages : "The gentleman who wished to translate my 'Origin of Species' has failed in getting a publisher. Baillière, Masson, and Hachette all rejected it with contempt. It was foolish and presumptuous in me, hoping to appear in a French dress ; but the idea would not have entered my head had it not been suggested to me. It is a great loss. I must console myself with the German edition which Prof. Bronn is bringing out." *

A sentence in another letter to M. de Quatrefages shows how anxious he was to convert one of the greatest of contemporary Zoologists : "How I should like to know whether Milne-Edwards has read the copy which I sent him, and whether he thinks I have made a pretty good case on our side of the question. There is no naturalist in the world for whose opinion I have so profound a respect. Of course I am not so silly as to expect to change his opinion."]

C. Darwin to C. Lyell.

Ilkley, [November 25th, 1859].

MY DEAR LYELL,—I have received your letter of the 24th. It is no use trying to thank you ; your kindness is beyond thanks. I will certainly leave out the whale and bear . . .

The edition was 1250 copies. When I was in spirits, I sometimes fancied that my book would be successful, but I never even built a castle in the air of such success as it has met with ; I do not mean the sale, but the impression it has made on you (whom I have always looked at as chief judge) and Hooker and Huxley. The whole has infinitely exceeded my wildest hopes.

Farewell, I am tired, for I have been going over the sheets.

My kind friend, farewell, yours,

C. DARWIN.

* See letters to Bronn, p. 276.

C. Darwin to C. Lyell.

Ilkley, Yorkshire,
December 2nd [1859].

MY DEAR LYELL,—Every note which you have sent me has'
interested me much.   Pray thank Lady Lyell for her remark.
In the chapters she refers to, I was unable to modify the passage in accordance to your suggestion ; but in the final
chapter I have modified three or four.   Kingsley, in a note *
to me, had a capital paragraph on such notions as mine being
not opposed to a high conception of the Deity.   I have inserted
it as an extract from a letter to me from a celebrated author
and divine.   I have put in about nascent organs.   I had the
greatest difficulty in partially making out Sedgwick's letter, and
I dare say I did greatly underrate its clearness.   Do what I
could, I fear I shall be greatly abused.   In answer to Sedgwick's remark that my book would be " mischievous," I asked
him whether truth can be known except by being victorious
over all attacks.   But it is no use.   H. C. Watson tells me
that one zoologist says he will read my book, " but I will never
believe it."   What a spirit to read any book in !   Crawford
writes to me that his notice † will be hostile, but that "he will
not calumniate the author."   He says he has read my book,
" at least such parts as he could understand."   He sent me
some notes and suggestions (quite unimportant), and they
show me that I have unavoidably done harm to the subject,
by publishing an abstract.   He is a real Pallasian ; nearly all
our domestic races descended from a multitude of wild species
now commingled.   I expected Murchison to be outrageous.

* The letter is given at Vol. II.
p. 287.

† John Crawford, orientalist, eth-
nologist, &c., b. 1783, d. 1868. The
review appeared in the *Examiner*,
and, though hostile, is free from
bigotry, as the following citation
will show : " We cannot help saying
that piety must be fastidious indeed
that objects to a theory the tendency of which is to show that all
organic beings, man included, are
in a perpetual progress of ameliora-
tion, and that is expounded in the
reverential language which we have
quoted."

How little he could ever have grappled with the subject
of denudation! How singular so great a geologist should
have so unphilosophical a mind! I have had several notes
from ——, very civil and less decided. Says he shall not
pronounce against me without much reflection, *perhaps will
say nothing* on the subject. X. says he will go to that part
of hell, which Dante tells us is appointed for those who are
neither on God's side nor on that of the devil.

I fully believe that I owe the comfort of the next few years
of my life to your generous support, and that of a very few
others. I do not think I am brave enough to have stood
being odious without support; now I feel as bold as a lion.
But there is one thing I can see I must learn, viz. to think
less of myself and my book. Farewell, with cordial thanks,

Yours most truly,
C. DARWIN.

I return home on the 7th, and shall sleep at Erasmus's. I
will call on you about ten o'clock, on Thursday, the 8th, and sit
with you, as I have so often sat, during your breakfast.

[In December there appeared in 'Macmillan's Magazine'
an article, "Time and Life," by Professor Huxley. It is
mainly occupied by an analysis of the argument of the
'Origin,' but it also gives the substance of a lecture deliver-
ed at the Royal Institution before that book was published.
Professor Huxley spoke strongly in favour of evolution in his
Lecture, and explains that in so doing he was to a great
extent resting on a knowledge of "the general tenor of the
researches in which Mr. Darwin had been so long engaged,"
and was supported in so doing by his perfect confidence in
his knowledge, perseverance, and "high-minded love of
truth." He was evidently deeply pleased by Mr. Huxley's
words, and wrote:

"I must thank you for your extremely kind notice of my
book in 'Macmillan.' No one could receive a more delightful

and honourable compliment. I had not heard of your Lecture, owing to my retired life. You attribute much too much to me from our mutual friendship. You have explained my leading idea with admirable clearness. What a gift you have of writing (or more properly thinking) clearly."]

C. Darwin to W. B. Carpenter.

Ilkley, Yorkshire,
December 3rd [1859].

MY DEAR CARPENTER,—I am perfectly delighted at your letter. It is a great thing to have got a great physiologist on our side. I say "our" for we are now a good and compact body of really good men, and mostly not old men. In the long-run we shall conquer. I do not like being abused, but I feel that I can now bear it; and, as I told Lyell, I am well convinced that it is the first offender who reaps the rich harvest of abuse. You have done an essential kindness in checking the odium theologicum in the [?].* It much pains all one's female relations and injures the cause.

I look at it as immaterial whether we go quite the same lengths ; and I suspect, judging from myself, that you will go further, by thinking of a population of forms like Ornitho- rhynchus, and by thinking of the common homological and embryological structure of the several vertebrate orders. But this is immaterial. I quite agree that the principle is every- thing. In my fuller MS. I have discussed a good many instincts ; but there will surely be more unfilled gaps here than with corporeal structure, for we have no fossil instincts, and know scarcely any except of European animals. When I reflect how very slowly I came round myself, I am in truth astonished at the candour shown by Lyell, Hooker, Huxley,

* This must refer to Carpenter's critique, which would now have been ready to appear in the January number of the 'National Review,' 1860, and in which the odium theo- logicum is referred to.

and yourself. In my opinion it is grand. I thank you cordially for taking the trouble of writing a review for the 'National.' God knows I shall have few enough in any degree favourable.*

C. Darwin to C. Lyell.

Saturday [December 5th, 1859].

. . . I have had a letter from Carpenter this morning. He reviews me in the 'National.' He is a convert, but does not go quite so far as I, but quite far enough, for he admits that all birds are from one progenitor, and probably all fishes and reptiles from another parent. But the last mouthful chokes him. He can hardly admit all vertebrates from one parent. He will surely come to this from Homology and Embryology. I look at it as grand having brought round a great physiologist, for great I think he certainly is in that line. How curious I shall be to know what line Owen will take: dead against us, I fear; but he wrote me a most liberal note on the reception of my book, and said he was quite prepared to consider fairly and without prejudice my line of argument.

C. Darwin to C. Lyell.

Down, Saturday [December 12th, 1859].

. . . I had very long interviews with ——, which perhaps you would like to hear about. . . . I infer from several expressions that, at bottom, he goes an immense way with us. . . . .

He said to the effect that my explanation was the best ever published of the manner of formation of species. I said I was very glad to hear it. He took me up short: "You must not at all suppose that I agree with you in all respects." I said I thought it no more likely that I should be right in

* See a letter to Dr. Carpenter, Vol. II. p. 262.

nearly all points, than that I should toss up a penny and get heads twenty times running. I asked him what he thought the weakest part. He said he had no particular objection to any part. He added :—

"If I must criticise, I should say, we do not want to know what Darwin believes and is convinced of, but what he can prove." I agreed most fully and truly that I have probably greatly sinned in this line, and defended my general line of argument of inventing a theory and seeing how many classes of facts the theory would explain. I added that I would endeavour to modify the "believes" and "convinceds." He took me up short : "You will then spoil your book, the charm of (!) it is that it is Darwin himself." He added another objection, that the book was too *teres atque rotundus*—that it explained everything, and that it was improbable in the highest degree that I should succeed in this. I quite agree with this rather queer objection, and it comes to this that my book must be very bad or very good. . . .

I have heard, by a roundabout channel, that Herschel says my book "is the law of higgledy-piggledy." What this exactly means I do not know, but it is evidently very contemptuous. If true this is a great blow and discouragement.

C. Darwin to John Lubbock.

December 14th [1859].

. . . The latter part of my stay at Ilkley did me much good, but I suppose I never shall be strong, for the work I have had since I came back has knocked me up a little more than once. I have been busy in getting a reprint (with a very few corrections) through the press.

My book has been as yet *very much* more successful than I ever dreamed of : Murray is now printing 3000 copies. Have you finished it ? If so, pray tell me whether you are with me on the *general* issue, or against me. If you are against me, I know well how honourable, fair, and candid an opponent I shall have, and which is a good deal more than I can say of all my opponents. . . .

Pray tell me what you have been doing. Have you had time for any Natural History? . . .

P.S.—I have got—I wish and hope I might say that *we* have got—a fair number of excellent men on our side of the question on the mutability of species.

J. D. Hooker to C. Darwin.

Kew [1859].

DEAR DARWIN,—You have, I know, been drenched with letters since the publication of your book, and I have hence forborne to add my mite.* I hope now that you are well through Edition II., and I have heard that you were flourishing in London. I have not yet got half-through the book, not from want of will, but of time—for it is the very hardest book to read, to full profits, that I ever tried—it is so cram-full of matter and reasoning. I am all the more glad that you have published in this form, for the three volumes, unprefaced by this, would have choked any Naturalist of the nineteenth century, and certainly have softened my brain in the operation of assimilating their contents. I am perfectly tired of marvelling at the wonderful amount of facts you have brought to bear, and your skill in marshalling them and throwing them on the enemy; it is also extremely clear as far as I have gone, but very hard to fully appreciate. Some-how it reads very different from the MS., and I often fancy that I must have been very stupid not to have more fully followed it in MS. Lyell told me of his criticisms. I did not appreciate them all, and there are many little matters I hope one day to talk over with you. I saw a highly flattering notice

* See, however, Vol. II. p. 228.

in the 'English Churchman,' short and not at all entering into discussion, but praising you and your book, and talking patronizingly of the doctrine! . . . Bentham and Henslow will still shake their heads, I fancy. . . .

Ever yours affectionately,

JOS. D. HOOKER.

C. Darwin to J. D. Hooker.

Down, December 14th [1859].

MY DEAR HOOKER,—Your approval of my book, for many reasons, gives me intense satisfaction ; but I must make some allowance for your kindness and sympathy. Any one with ordinary faculties, if he had *patience* enough and plenty of time, could have written my book. You do not know how I admire your and Lyell's generous and unselfish sympathy ; I do not believe either of you would have cared so much about your own work. My book, as yet, has been far more successful than I ever even formerly ventured in the wildest daydreams to anticipate. We shall soon be a good body of working men, and shall have, I am convinced, all young and rising naturalists on our side. I shall be intensely interested to hear whether my book produces any effect on A. Gray ; from what I heard at Lyell's, I fancy your correspondence has brought him some way already. I fear that there is no chance of Bentham being staggered. Will he read my book ? Has he a copy ? I would send him one of the reprints if he has not. Old J. E. Gray,* at the British Museum, attacked me in fine style : "You have just reproduced Lamarck's doc-

* John Edward Gray (born 1800, died 1875) was the son of S. F. Gray, author of the 'Supplement to the Pharmacopœia.' In 1821 he published in his father's name 'The Natural Arrangement of British Plants,' one of the earliest works in English on the natural method. In 1824 he became connected with the Natural History Department of the British Museum, and was appointed Keeper of the Zoological collections in 1840. He was the author of 'Illustrations of Indian Zoology,' 'The Knowsley Menagerie,' &c., and of innumerable descriptive Zoological papers.

trine, and nothing else, and here Lyell and others have been attacking him for twenty years, and because *you* (with a sneer and laugh) say the very same thing, they are all coming round ; it is the most ridiculous inconsistency, &c. &c."

You must be very glad to be settled in your house, and I hope all the improvements satisfy you. As far as my experience goes, improvements are never perfection. I am very sorry to hear that you are still so very busy, and have so much work. And now for the main purport of my note, which is to ask and beg you and Mrs. Hooker (whom it is really an age since I have seen), and all your children, if you like, to come and spend a week here. It would be a great pleasure to me and to my wife. . . . As far as we can see, we shall be at home all the winter ; and all times probably would be equally convenient ; but if you can, do not put it off very late, as it may slip through. Think of this and persuade Mrs. Hooker, and be a good man and come.

Farewell, my kind and dear friend,
Yours affectionately,
C. DARWIN.

P.S.—I shall be very curious to hear what you think of my discussion on Classification in Chap. XIII. ; I believe Huxley demurs to the whole, and says he has nailed his colours to the mast, and I would sooner die than give up ; so that we are in as fine a frame of mind to discuss the point as any two religionists.

Embryology is my pet bit in my book, and, confound my friends, not one has noticed this to me.

C. Darwin to Asa Gray.

Down, December 21st [1859].

MY DEAR GRAY,—I have just received your most kind, long, and valuable letter. I will write again in a few days, for I am at present unwell and much pressed with business :

to-day's note is merely personal. I should, for several reasons, be very glad of an American Edition. I have made up my mind to be well abused ; but I think it of importance that my notions should be read by intelligent men, accustomed to scientific argument, though *not* naturalists. It may seem absurd, but I think such men will drag after them those naturalists who have too firmly fixed in their heads that a species is an entity. The first edition of 1250 copies was sold on the first day, and now my publisher is printing off, as *rapidly as possible*, 3000 more copies. I mention this solely because it renders probable a remunerative sale in America. I should be infinitely obliged if you could aid an American reprint ; and could make, for my sake and the publisher's, any arrangement for any profit. The new edition is only a reprint, yet I have made a *few* important corrections.   I will have the clean sheets sent over in a few days of as many sheets as are printed off, and the remainder afterwards, and you can do anything you like,—if nothing, there is no harm done. I should be glad for the new edition to be reprinted and not the old.—In great haste, and with hearty thanks,

Yours very sincerely,
C. DARWIN.

I will write soon again.

C. Darwin to C. Lyell.

Down, 22nd [December, 1859].

MY DEAR LYELL,—Thanks about " Bears," * a word of ill-omen to me.

I am too unwell to leave home, so shall not see you.

I am very glad of your remarks on Hooker.† I have not yet

---

* See ' Origin,' ed. i., p. 184.

† Sir C. Lyell wrote to Sir J. D. Hooker, Dec. 19, 1859 (' Life,' ii. p. 327) : " I have just finished the reading of your splendid Essay [the ' Flora of Australia '] on the origin of species, as illustrated by your wide botanical experience, and think it goes very far to raise the variety-making hypothesis to the rank of a theory, as accounting for the manner in which new species enter the world."

got the Essay. The parts which I read in sheets seemed to me grand, especially the generalization about the Australian flora itself. How superior to Robert Brown's celebrated essay! I have not seen Naudin's paper,* and shall not be able till I hunt the libraries. I am very anxious to see it. Decaisne seems to think he gives my whole theory. I do not know when I shall have time and strength to grapple with Hooker. . . .

P.S.—I have heard from Sir W. Jardine :† his criticisms are quite unimportant ; some of the Galapagos so-called species ought to be called varieties, which I fully expected ; some of the sub-genera, thought to be wholly endemic, have been found on the Continent (not that he gives his authority), but I do not make out that the species are the same. His letter is brief and vague, but he says he will write again.

C. Darwin to J. D. Hooker.

Down [23rd December, 1859].

MY DEAR HOOKER,—I received last night your 'Introduction,' for which very many thanks ; I am surprised to see

---

* 'Revue Horticole,' 1852. See Historical Sketch in the later editions of the 'Origin of Species.'

† Jardine, Sir William, Bart., b. 1800, d. 1874, was the son of Sir A. Jardine of Applegarth, Dumfriesshire. He was educated at Edinburgh, and succeeded to the title on his father's decease in 1821. He published, jointly with Mr. Prideaux J. Selby, Sir Stamford Raffles, Dr. Horsfield, and other ornithologists, ' Illustrations of Ornithology,' and edited the 'Naturalist's Library,' in 40 vols. which included the four branches : Mammalia, Ornithology, Ichthyology, and Entomology. Of these 40 vols. 14 were written by himself. In 1836 he became editor of the 'Magazine of Zoology and Botany,' which, two years later, was transformed into ' Annals of Natural History,' but remained under his direction. For Bohn's Standard Library he edited White's ' Natural History of Selborne.' Sir W. Jardine was also joint editor of the ' Edinburgh Philosophical Journal,' and was author of ' British Salmonidae,' ' Ichthyology of Annandale,' ' Memoirs of the late Hugh Strickland,' ' Contributions to Ornithology,' ' Ornithological Synonyms,' &c. —(Taken from Ward, ' Men of the Reign,' and Cates, ' Dictionary of General Biography.')

how big it is : I shall not be able to read it very soon.   It
was very good of you to send Naudin, for I was very curious
to see it.   I am surprised that Decaisne should say it was
the same as mine.   Naudin gives artificial selection, as well
as a score of English writers, and when he says species were
formed in the same manner, I thought the paper would certainly prove exactly the same as mine.   But I cannot find
one word like the struggle for existence and natural selection.
On the contrary, he brings in his principle (p. 103) of finality
(which I do not understand), which, he says, with some authors
is fatality, with others providence, and which adapts the forms
of every being, and harmonises them all throughout nature.

He assumes like old geologists (who assumed that the forces
of nature were formerly greater), that species were at first
more plastic.   His simile of tree and classification is like
mine (and others), but he cannot, I think, have reflected
much on the subject, otherwise he would see that genealogy
by itself does not give classification ; I declare I cannot see a
much closer approach to Wallace and me in Naudin than
in Lamarck—we all agree in modification and descent.   If
I do not hear from you I will return the ' Revue ' in a few
days (with the cover).   I dare say Lyell would be glad to see
it.   By the way, I will retain the volume till I hear whether
I shall or not send it to Lyell.   I should rather like Lyell
to see this note, though it is foolish work sticking up for
independence or priority.

Ever yours,

C. DARWIN.

A. Sedgwick* to C. Darwin.

Cambridge, December 24th, 1859.

MY DEAR DARWIN,—I write to thank you for your work on
the ' Origin of Species.'   It came, I think, in the latter part

* Rev. Adam Sedgwick, Wood-   the University of Cambridge. Born
wardian Professor of Geology in   1785, died 1873.

of last week; but it *may* have come a few days sooner, and
been overlooked among my book-parcels, which often remain
unopened when I am lazy or busy with any work before me.
So soon as I opened it I began to read it, and I finished it,
after many interruptions, on Tuesday.   Yesterday I was employed—1st, in preparing for my lecture; 2ndly, in attending
a meeting of my brother Fellows to discuss the final propositions of the Parliamentary Commissioners; 3rdly, in lecturing;
4thly, in hearing the conclusion of the discussion and the
College reply, whereby, in conformity with my own wishes, we
accepted the scheme of the Commissioners; 5thly, in dining
with an old friend at Clare College; 6thly, in adjourning to
the weekly meeting of the Ray Club, from which I returned
at 10 P.M., dog-tired, and hardly able to climb my staircase.
Lastly, in looking through the *Times* to see what was going
on in the busy world.

I do not state this to fill space (though I believe that
Nature does abhor a vacuum), but to prove that my reply and
my thanks are sent to you by the earliest leisure I have, though
that is but a very contracted opportunity.   If I did not think
you a good-tempered and truth-loving man, I should not tell
you that (spite of the great knowledge, store of facts, capital
views of the correlation of the various parts of organic nature,
admirable hints about the diffusion, through wide regions, of
many related organic beings, &c. &c.) I have read your book
with more pain than pleasure.   Parts of it I admired greatly,
parts I laughed at till my sides were almost sore; other parts
I read with absolute sorrow, because I think them utterly
false and grievously mischievous.   You have *deserted*—after
a start in that tram-road of all solid physical truth—the true
method of induction, and started us in machinery as wild,
I think, as Bishop Wilkins's locomotive that was to sail with
us to the moon.   Many of your wide conclusions are based
upon assumptions which can neither be proved nor disproved,
why then express them in the language and arrangement

of philosophical induction? As to your grand principle—
natural selection—what is it but a secondary consequence of
supposed, or known, primary facts? Development is a better
word, because more close to the cause of the fact? For you
do not deny causation. I call (in the abstract) causation the
will of God; and I can prove that He acts for the good of
His creatures. He also acts by laws which we can study
and comprehend. Acting by law, and under what is called
final causes, comprehends, I think, your whole principle.
You write of "natural selection" as if it were done consciously
by the selecting agent. 'Tis but a consequence of the pre-
supposed development, and the subsequent battle for life.
This view of nature you have stated admirably, though
admitted by all naturalists and denied by no one of common
sense. We all admit development as a fact of history: but
how came it about? Here, in language, and still more in
logic, we are point-blank at issue. There is a moral or metaphysical part of nature as well as a physical. A man who
denies this is deep in the mire of folly. 'Tis the crown and
glory of organic science that it *does* through *final cause*, link
material and moral; and yet *does not* allow us to mingle
them in our first conception of laws, and our classification
of such laws, whether we consider one side of nature or the
other. You have ignored this link; and, if I do not mistake
your meaning, you have done your best in one or two preg-
nant cases to break it. Were it possible (which, thank God, it is
not) to break it, humanity, in my mind, would suffer a damage
that might brutalize it, and sink the human race into a lower
grade of degradation than any into which it has fallen since
its written records tell us of its history. Take the case of the
bee-cells. If your development produced the successive
modification of the bee and its cells (which no mortal can
prove), final cause would stand good as the directing cause
under which the successive generations acted and gradually
improved. Passages in your book, like that to which I have

alluded (and there are others almost as bad), greatly shocked my moral taste. I think, in speculating on organic descent, you *over*-state the evidence of geology ; and that you *under*-state it while you are talking of the broken links of your natural pedigree : but my paper is nearly done, and I must go to my lecture-room. Lastly, then, I greatly dislike the concluding chapter—not as a summary, for in that light it appears. good—but I dislike it from the tone of triumphant confidence in which you appeal to the rising generation (in a tone I condemned in the author of the 'Vestiges') and prophecy of things. not yet in the womb of time, nor (if we are to trust the accumulated experience of human sense and the inferences of its logic) ever likely to be found anywhere but in the fertile womb of man's imagination. And now to say a word about a son of a monkey and an old friend of yours : I am better, far better, than I was last year. I have been lecturing three days a week (formerly I gave six a week) without much fatigue, but I find by the loss of activity and memory, and of all productive powers, that my bodily frame is sinking slowly towards the earth. But I have visions of the future. They are as much a part of myself as my stomach and my heart, and these visions are to have their antitype in solid fruition of what is best and greatest. But on one condition only—that I humbly accept God's revelation of Himself both in His works and in. His word, and do my best to act in conformity with that knowledge which He only can give me, and He only can sustain me in doing. If you and I do all this, we shall meet. in heaven.

I have written in a hurry, and in a spirit of brotherly love,. therefore forgive any sentence you happen to dislike ; and believe me, spite of any disagreement in some points of the deepest moral interest, your true-hearted old friend,

A. SEDGWICK.

C. Darwin to T. H. Huxley.

Down, Dec. 25th [1859].

MY DEAR HUXLEY,—One part of your note has pleased me so much that I must thank you for it. Not only Sir H. H. [Holland], but several others, have attacked me about analogy leading to belief in one primordial *created* form.* (By which I mean only that we know nothing as yet [of] how life originates.) I thought I was universally condemned on this head. But I answered that though perhaps it would have been more prudent not to have put it in, I would not strike it out, as it seemed to me probable, and I give it on no other grounds. You will see in your mind the kind of arguments which made me think it probable, and no one fact had so great an effect on me as your most curious remarks on the apparent homologies of the head of Vertebrata and Articulata.

You have done a real good turn in the Agency business † (I never before heard of a hard-working, unpaid agent besides yourself), in talking with Sir H. H., for he will have great influence over many. He floored me from my ignorance about the bones of the ear, and I made a mental note to ask you what the facts were.

With hearty thanks and real admiration for your generous zeal for the subject.

Yours most truly,

C. DARWIN.

You may smile about the care and precautions I have taken about my ugly MS. ; ‡ it is not so much the value I set on

---

* 'Origin,' edit. i. p. 484.— "Therefore I should infer from analogy that probably all the organic beings which have ever lived on this earth have descended from some one primordial form, into which life was first breathed."

† "My General Agent" was a sobriquet applied at this time by my father to Mr. Huxley.

‡ Manuscript left with Mr. Huxley for his perusal.

them, but the remembrance of the intolerable labour—for instance, in tracing the history of the breeds of pigeons.

C. Darwin to J. D. Hooker.

Down, 25th [December, 1859].

. . . I shall not write to Decaisne ; * I have always had a strong feeling that no one had better defend his own priority.   I cannot say that I am as indifferent to the subject as I ought to be, but one can avoid doing anything in consequence.

I do not believe one iota about your having assimilated any of my notions unconsciously.   You have always done me more than justice.   But I do think I did you a bad turn by getting you to read the old MS., as it must have checked your own original thoughts.   There is one thing I am fully convinced of, that the future progress (which is the really important point) of the subject will have depended on really good and well-known workers, like yourself, Lyell, and Huxley, having taken up the subject, than on my own work.   I see plainly it is this that strikes my non-scientific friends.

Last night I said to myself, I would just cut your Introduction, but would not begin to read, but I broke down, and had a good hour's read.

Farewell, yours affectionately,

C. DARWIN.

C. Darwin to J. D. Hooker.

December 28th, 1859.

. . . Have you seen the splendid essay and notice of my book in the *Times ?* †   I cannot avoid a strong suspicion that it is by Huxley ; but I never heard that he wrote in the *Times.*   It will do grand service, . . .

* With regard to Naudin's paper in the ' Revue Horticole,' 1852.
† Dec. 26th.

C. Darwin to T. H. Huxley.

Down, Dec. 28th [1859].

MY DEAR HUXLEY,—Yesterday evening, when I read the *Times* of a previous day, I was amazed to find a splendid essay and review of me. Who can the author be? I am intensely curious. It included an eulogium of me which quite touched me, though I am not vain enough to think it all deserved. The author is a literary man, and German scholar. He has read my book very attentively; but, what is very remarkable, it seems that he is a profound naturalist. He knows my Barnacle-book, and appreciates it too highly. Lastly, he writes and thinks with quite uncommon force and clearness; and what is even still rarer, his writing is seasoned with most pleasant wit. We all laughed heartily over some of the sentences. I was charmed with those unreasonable mortals, who know anything, all thinking fit to range themselves on one side.* Who can it be? Certainly I should have said that there was only one man in England who could have written this essay, and that *you* were the man. But I suppose I am wrong, and that there is some hidden genius of great calibre. For how could you influence Jupiter Olympius and make him give three and a half columns to pure science? The old fogies will think the world will come to an end. Well, whoever the man is, he has done great service to the cause, far more than by a dozen reviews in common periodicals. The grand way he soars above common religious

---

* The reviewer proposes to pass by the orthodox view, according to which the phenomena of the organic world are "the immediate product of a creative fiat, and consequently are out of the domain of science altogether." And he does so "with less hesitation, as it so happens that those persons who are prac- tically conversant with the facts of the case (plainly a considerable advantage) have always thought fit to range themselves" in the category of those holding "views which profess to rest on a scientific basis only, and therefore admit of being argued to their conse- quences."

prejudices, and the admission of such views into the *Times*, I look at as of the highest importance, quite independently of the mere question of species. If you should happen to be *acquainted* with the author, for Heaven-sake tell me who he is?

My dear Huxley, yours most sincerely,

C. DARWIN.

[It is impossible to give in a short space an adequate idea of Mr. Huxley's article in the *Times* of December 26. It is admirably planned, so as to claim for the 'Origin' a respectful hearing, and it abstains from anything like dogmatism in asserting the truth of the doctrines therein upheld. A few passages may be quoted :—" That this most ingenious hypothesis enables us to give a reason for many apparent anomalies in the distribution of living beings in time and space, and that it is not contradicted by the main phenomena of life and organisation, appear to us to be unquestionable." Mr. Huxley goes on to recommend to the readers of the 'Origin' a condition of "*thätige Skepsis*"—a state of "doubt which so loves truth that it neither dares rest in doubting, nor extinguish itself by unjustified belief." The final paragraph is in a strong contrast to Professor Sedgwick and his "ropes of bubbles" (see p. 298). Mr. Huxley writes : " Mr. Darwin abhors mere speculation as nature abhors a vacuum. He is as greedy of cases and precedents as any constitutional lawyer, and all the principles he lays down are capable of being brought to the test of observation and experiment. The path he bids us follow professes to be not a mere airy track, fabricated of ideal cobwebs, but a solid and broad bridge of facts. If it be so, it will carry us safely over many a chasm in our knowledge, and lead us to a region free from the snares of those fascinating but barren virgins, the Final Causes, against whom a high authority has so justly warned us."

There can be no doubt that this powerful essay, appearing

as it did in the leading daily Journal, must have had a strong influence on the reading public. Mr. Huxley allows me to quote from a letter an account of the happy chance that threw into his hands the opportunity of writing it.

"The 'Origin' was sent to Mr. Lucas, one of the staff of the *Times* writers at that day, in what I suppose was the ordinary course of business. Mr. Lucas, though an excellent journalist, and, at a later period, editor of 'Once a Week,' was as innocent of any knowledge of science as a babe, and bewailed himself to an acquaintance on having to deal with such a book. Whereupon he was recommended to ask me to get him out of his difficulty, and he applied to me accordingly, explaining, however, that it would be necessary for him formally to adopt anything I might be disposed to write, by prefacing it with two or three paragraphs of his own.

"I was too anxious to seize upon the opportunity thus offered of giving the book a fair chance with the multitudinous readers of the *Times* to make any difficulty about conditions; and being then very full of the subject, I wrote the article faster, I think, than I ever wrote anything in my life, and sent it to Mr. Lucas, who duly prefixed his opening sentences.

"When the article appeared, there was much speculation as to its authorship. The secret leaked out in time, as all secrets will, but not by my aid; and then I used to derive a good deal of innocent amusement from the vehement assertions of some of my more acute friends, that they knew it was mine from the first paragraph!

"As the *Times* some years since, referred to my connection with the review, I suppose there will be no breach of confidence in the publication of this little history, if you think it worth the space it will occupy."]

CHAPTER VII.

THE 'ORIGIN OF SPECIES'—(*continued*).

1860.

I EXTRACT a few entries from my father's Diary :—

"Jan. 7th. The second edition, 3000 copies, of 'Origin' was published."

"May 22nd. The first edition of 'Origin' in the United States was 2500 copies."

My father has here noted down the sums received for the Origin.'

First Edition	..	..	..£180 0 0
Second Edition	..	..	.. 636 13 4
			£816 13 4

After the publication of the second edition he began at once, on Jan. 9th, looking over his materials for the 'Variation of Animals and Plants ;' the only other work of the year was on Drosera.

He was at Down during the whole of this year, except for a visit to Dr. Lane's Water-cure Establishment at Sudbrooke, in June, and for visits to Miss Elizabeth Wedgwood's house at Hartfield, in Sussex (July), and to Eastbourne, Sept. 22 to Nov. 16.

C. Darwin to J. D. Hooker.

Down, January 3rd [1860].

MY DEAR HOOKER,—I have finished your Essay.* As probably you would like to hear my opinion, though a non-botanist, I will give it without any exaggeration. To my judgment it is by far the grandest and most interesting essay, on subjects of the nature discussed, I have ever read. You know how I admired your former essays, but this seems to me far grander. I like all the part after p. xxvi better than the first part, probably because newer to me. I dare say you will demur to this, for I think every author likes the most speculative parts of his own productions. How superior your essay is to the famous one of Brown (here will be sneer 1st from you). You have made all your conclusions so admirably clear, that it would be no use at all to be a botanist (sneer No. 2). By Jove, it would do harm to affix any idea to the long names of outlandish orders. One can look at your con-clusions with the philosophic abstraction with which a mathe-matician looks at his $a \times x + \sqrt{z^2}$, &c. &c. I hardly know which parts have interested me most ; for over and over again I exclaimed, "this beats all." The general comparison of the Flora of Australia with the rest of the world, strikes me (as before) as extremely original, good, and suggestive of many reflections.

. . . . The invading Indian Flora is very interesting, but I think the fact you mention towards the close of the essay— that the Indian vegetation, in contradistinction to the Ma-layan vegetation, is found in low and level parts of the Malay Islands, *greatly* lessens the difficulty which at first (page 1) seemed so great. There is nothing like one's own hobby-horse. I suspect it is the same case as of glacial migration, and of naturalised production—of production of greater area

* 'Australian Flora.

conquering those of lesser; of course the Indian forms would have a greater difficulty in seizing on the cool parts of Australia. I demur to your remarks (page 1), as not "conceiving anything in soil, climate, or vegetation of India," which could stop the introduction of Australian plants. Towards the close of the essay (page civ), you have admirable remarks on our profound ignorance of the cause of possible naturalisation or introduction; I would answer p. 1, by a later page, viz. p. civ.

Your contrast of the south-west and south-east corners is one of the most wonderful cases I ever heard of. . . . You show the case with wonderful force. Your discussion on mixed invaders of the south-east corner (and of New Zealand) is as curious and intricate a problem as of the races of men in Britain. Your remark on a mixed invading Flora keeping down or destroying an original Flora, which was richer in number of species, strikes me as *eminently new and important.* I am not sure whether to me the discussion on the New Zealand Flora is not even more instructive. I cannot too much admire both. But it will require a long time to suck in all the facts. Your case of the largest Australian orders having none, or very few, species in New Zealand, is truly marvellous. Anyhow, you have now *demonstrated* (together with no mammals in New Zealand) (bitter sneer No. 3), that New Zealand has never been continuously, or even nearly continuously, united by land to Australia!! At p. lxxxix, is the only sentence (on this subject) in the whole essay at which I am much inclined to quarrel, viz. that no theory of trans-oceanic migration can explain, &c. &c. Now I maintain against all the world, that no man knows anything about the trans-oceanic power of migration. You do not know whether or not the absent orders have seeds which are killed by sea-water, like almost all Leguminosæ, and like another order which I forget. Birds do not migrate from Australia to New Zealand, and therefore floatation *seems* the only possible means ; but yet I maintain that we do not know enough to argue on the question, especially as we do not know the main fact whether the seeds of Australian orders are killed by sea-water.

The discussion on European Genera is profoundly interesting ; but here alone I earnestly beg for more information, viz. to know which of these genera are absent in the Tropics of the world, *i.e.* confined to temperate regions. I excessively wish to know, *on the notion of Glacial Migration,* how much modification has taken place in Australia. I had better explain when we meet, and get you to go over and mark the list.

. . . . The list of naturalised plants is extremely interesting, but why at the end, in the name of all that is good and bad, do you not sum up and comment on your facts ? Come, I will have a sneer at you in return for the many which you will have launched at this letter. Should you [not] have remarked on the number of plants naturalised in Australia and the United States *under extremely different climates,* as showing that climate is so important, and [on] the considerable sprinkling of plants from India, North America, and South Africa, as showing that the frequent introduction of seeds is so important ? With respect to "abundance of unoccupied ground in Australia," do you believe that European plants introduced by man now grow on spots in Australia which were absolutely bare ? But I am an impudent dog, one must defend one's own fancy theories against such cruel men as you. I dare say this letter will appear very conceited, but one must form an opinion on what one reads with attention, and in simple truth, I cannot find words strong enough to express my admiration of your essay.

My dear old friend, yours affectionately,

C. DARWIN.

P.S.—I differ about the *Saturday Review*.* One cannot expect fairness in a reviewer, so I do not complain of all the other arguments besides the 'Geological Record' being omitted. Some of the remarks about the lapse of years are very good, and the reviewer gives me some good and well-deserved raps—confound it. I am sorry to confess the truth : but it does not at all concern the main argument. That was a nice notice in the *Gardeners' Chronicle*. I hope and imagine that Lindley is almost a convert. Do not forget to tell me if Bentham gets at all more staggered.

With respect to tropical plants during the Glacial period, I throw in your teeth your own facts, at the base of the Himalaya, on the possibility of the co-existence of at least forms of the tropical and temperate regions. I can give a parallel case for animals in Mexico. Oh! my dearly beloved puny child, how cruel men are to you! I am very glad you approve of the Geographical chapters. . . .

C. Darwin to C. Lyell.

Down [January 4th, 1860].

MY DEAR L.—*Gardeners' Chronicle* returned safe. Thanks for note. I am beyond measure glad that you get more and more roused on the subject of species, for, as I have always said, I am well convinced that your opinions and writings will do far more to convince the world than mine. You will make a grand discussion on man. You are very bold in this, and I honour you. I have been, like you, quite surprised at the want of originality in opposed arguments and in favour too. Gwyn Jeffreys attacks me justly in his letter about strictly littoral shells not being often embedded at least

* *Saturday Review*, Dec. 24, 1859. The hostile arguments of the reviewer are geological, and he deals especially with the denudation of the Weald. The reviewer remarks that, "if a million of centuries, more or less, is needed for any part of his argument, he feels no scruple in taking them to suit his purpose."

in Tertiary deposits.   I was in a muddle, for I was thinking
of Secondary, yet Chthamalus applied to Tertiary. . . . . .

Possibly you might like to see the enclosed note * from
Whewell, merely as showing that he is not horrified with us.
You can return it whenever you have occasion to write, so as
not to waste your time.

C. D.

C. Darwin to C. Lyell.

Down [January 4th ? 1860].

. . . . . . I have had a brief note from Keyserling,† but not
worth sending you.   He believes in change of species, grants
that natural selection explains well adaptation of form, but
thinks species change too regularly, as if by some chemical
law, for natural selection to be the sole cause of change.
I can hardly understand his brief note, but this is I think
the upshot.

. . . . . I will send A. Murray's paper whenever published.‡

---

* Dr. Whewell wrote (Jan. 2, 1860) : " . . . I cannot, yet at least, become a convert.   But there is so much of thought and of fact in what you have written that it is not to be contradicted without careful selection of the ground and manner of the dissent."   Dr. Whewell dissented in a practical manner for some years, by refusing to allow a copy of the ' Origin of Species ' to be placed in the Library of Trinity College.

† Count Keyserling, geologist, joint author with Murchison of the ' Geology of Russia,' 1845 ; and mentioned in Prof. Geikie's ' Life of Murchison.'

‡ The late Andrew Murray wrote two papers on the ' Origin ' in the Proc. R. Soc. Edin. 1860. The one referred to here is dated Jan. 16, 1860. The following is quoted from p. 6 of the separate copy : " But the second, and, as it appears to me, by much the most important phase of reversion to type (and which is practically, if not altogether ignored by Mr. Darwin), is the instinctive inclination which induces individuals of the same species by preference to inter-cross with those possessing the qualities which they themselves want, so as to preserve the purity or equilibrium of the breed. . . . It is trite to a proverb, that tall men marry little women . . . a man of genius marries a fool . . . and we are told that this is the result of the charm of contrast, or of qualities admired in others because we do not possess them.   I do not so explain it.   I imagine it is the effort of nature to preserve the typical medium of the race."

It includes speculations (which perhaps he will modify) so rash, and without a single fact in support, that had I advanced them he or other reviewers would have hit me very hard. I am sorry to say that I have no "consolatory view" on the dignity of man. I am content that man will probably advance, and care not much whether we are looked at as mere savages in a remotely distant future. Many thanks for your last note.

Yours affectionately,
C. DARWIN.

I have received, in a Manchester newspaper, rather a good squib, showing that I have proved "might is right," and therefore that Napoleon is right, and every cheating tradesman is also right.

C. Darwin to W. B. Carpenter.

Down, January 6th [1860]?

MY DEAR CARPENTER,—I have just read your excellent article in the 'National.' It will do great good ; especially if it becomes known as your production. It seems to me to give an excellently clear account of Mr. Wallace's and my views. How capitally you turn the flanks of the theological opposers by opposing to them such men as Bentham and the more philosophical of the systematists ! I thank you sincerely for the *extremely* honourable manner in which you mention me. I should have liked to have seen some criticisms or remarks on embryology, on which subject you are so well instructed. I do not think any candid person can read your article without being much impressed with it. The old doctrine of immutability of specific forms will surely but slowly die away. It is a shame to give you trouble, but I should be very much obliged if you could tell me where differently coloured eggs in individuals of the cuckoo have been described, and their laying in twenty-seven kinds of nests. Also do you know from your own observation that the limbs of sheep imported into the West Indies change colour? I have had detailed information about the loss of wool; but my accounts made the change slower than you describe.

With most cordial thanks and respect, believe me, my dear Carpenter, yours very sincerely,

CH. DARWIN.

C. Darwin to L. Jenyns. *

Down, January 7th, 1860.

MY DEAR JENYNS,—I am very much obliged for your letter. It is of great use and interest to me to know what impression my book produces on philosophical and instructed minds. I thank you for the kind things which you say; and you go with me much further than I expected. You will think it presumptuous, but I am convinced, *if circumstances lead you to keep the subject in mind,* that you will go further. No one has yet cast doubts on my explanation of the subordination of group to group, on homologies, embryology, and rudimentary organs; and if my explanation of these classes of facts be at all right, whole classes of organic beings must be included in one line of descent.

The imperfection of the Geological Record is one of the greatest difficulties. . . . . During the earliest period the record would be most imperfect, and this seems to me sufficiently to account for our not finding intermediate forms between the classes in the same great kingdoms. It was certainly rash in me putting in my belief of the probability of all beings having descended from *one* primordial form; but as this seems yet to me probable, I am not willing to strike it out. Huxley alone supports me in this, and something could be said in its favour. With respect to man, I am very far from wishing to obtrude my belief; but I thought it dishonest to quite conceal my opinion. Of course it is

* Rev. L. Blomefield.

open to every one to believe that man appeared by a
separate miracle, though I do not myself see the necessity or
probability.

Pray accept my sincere thanks for your kind note. Your
going some way with me gives me great confidence that I am
not very wrong. For a very long time I halted half-way; but
I do not believe that any enquiring mind will rest half-way.
People will have to reject all or admit all; by *all*, I mean
only the members of each great kingdom.

My dear Jenyns, yours most sincerely,
C. DARWIN.

C. Darwin to C. Lyell.

Down, January 10th [1860].

. . . It is perfectly true that I owe nearly all the corrections*
to you, and several verbal ones to you and others; I am
heartily glad you approve of them, as yet only two things
have annoyed me; those confounded millions† of years (not
that I think it is probably wrong), and my not having (by
inadvertence) mentioned Wallace towards the close of the
book in the summary, not that any one has noticed this to me.
I have now put in Wallace's name at p. 484 in a conspicuous
place. I cannot refer you to tables of mortality of children,
&c. &c. I have notes somewhere, but I have not the *least*
idea where to hunt, and my notes would now be old. I shall
be truly glad to read carefully any MS. on man, and give my
opinion. You used to caution me to be cautious about man.

---

* The second edition of 3000
copies of the 'Origin' was published on January 7th.

† This refers to the passage in
the 'Origin of Species' (2nd edit.
p. 285), in which the lapse of time
implied by the denudation of the
Weald is discussed. The discussion closes with the sentence: "So
that it is not improbable that a
longer period than 300 million
years has elapsed since the latter
part of the Secondary period."
This passage is omitted in the later
editions of the 'Origin,' against the
advice of some of his friends, as
appears from the pencil notes in
my father's copy of the 2nd edition.

I suspect I shall have to return the caution a hundred fold! Yours will, no doubt, be a grand discussion; but it will horrify the world at first more than my whole volume; although by the sentence (p. 489, new edition *) I show that I believe man is in the same predicament with other animals. It is in fact impossible to doubt it. I have thought (only vaguely) on man. With respect to the races, one of my best chances of truth has broken down from the impossibility of getting facts. I have one good speculative line, but a man must have entire credence in Natural Selection before he will even listen to it. Psychologically, I have done scarcely any-thing. Unless, indeed, expression of countenance can be included, and on that subject I have collected a good many facts, and speculated, but I do not suppose I shall ever publish, but it is an uncommonly curious subject. By the way I sent off a lot of questions the day before yesterday to Tierra del Fuego on expression! I suspect (for I have never read it) that Spencer's 'Psychology' has a bearing on Psychology as we should look at it. By all means read the Preface, in about 20 pages, of Hensleigh Wedgwood's new Dictionary, on the first origin of Language; Erasmus would lend it. I agree about Carpenter, a very good article, but with not much original. . . . Andrew Murray has criticised, in an address to the Botanical Society of Edinburgh, the notice in the 'Linnean Journal,' and "has disposed of" the whole theory by an ingenious difficulty, which I was very stupid not to have thought of; for I express surprise at more and analogous cases not being known. The difficulty is, that amongst the blind insects of the caves in distant parts of the world there are some of the same genus, and yet the genus is not found out of the caves or living in the free world. I have little doubt that, like the fish Amblyopsis, and like Proteus in Europe, these insects are "wrecks of ancient life," or "living fossils," saved from competition and extermination. But that

* First edition, p. 488.

formerly *seeing* insects of the same genus roamed over the whole area in which the cases are included.

Farewell, yours affectionately,

C. DARWIN.

P.S.—*Our* ancestor was an animal which breathed water, had a swim bladder, a great swimming tail, an imperfect skull, and undoubtedly was an hermaphrodite! Here is a pleasant genealogy for mankind.

C. Darwin to C. Lyell.

Down, January 14th [1860].

. . . I shall be much interested in reading your man discussion, and will give my opinion carefully, whatever that may be worth; but I have so long looked at you as the type of cautious scientific judgment (to my mind one of the highest and most useful qualities), that I suspect my opinion will be superfluous. It makes me laugh to think what a joke it will be if I have to caution you, after your cautions on the same subject to me!

I will order Owen's book;* I am very glad to hear Huxley's opinion on his classification of man; without having due knowledge, it seemed to me from the very first absurd; all classifications founded on single characters I believe have failed.

. . . What a grand immense benefit you conferred on me by getting Murray to publish my book. I never till to-day realised that it was getting widely distributed; for in a letter from a lady to-day to E., she says she heard a man enquiring for it at the *Railway Station!!!* at Waterloo Bridge; and the bookseller said that he had none till the new edition was out. The bookseller said he had not read it, but had heard it was a very remarkable book!!! . . . .

* 'Classification of the Mammalia,' 1859.

C. Darwin to J. D. Hooker.

Down, 14th [January, 1860].

. . . . I heard from Lyell this morning, and he tells
me a piece of news. You are a good-for-nothing man ; here
you are slaving yourself to death with hardly a minute to spare,
and you must write a review on my book ! I thought it * a
very good one, and was so much struck with it, that I sent it
to Lyell. But I assumed, as a matter of course, that it was
Lindley's. Now that I know it is yours, I have re-read it, and
my kind and good friend, it has warmed my heart with all the
honourable and noble things you say of me and it. I was a
good deal surprised at Lindley hitting on some of the remarks,
but I never dreamed of you. I admired it chiefly as so well
adapted to tell on the readers of the *Gardeners' Chronicle ;*
but now I admire it in another spirit. Farewell, with hearty
thanks. ; . . . Lyell is going at man with an audacity that
frightens me. It is a good joke ; he used always to caution
me to slip over man.

[In the *Gardeners' Chronicle*, Jan. 21, 1860, appeared a
short letter from my father, which was called forth by
Mr. Westwood's communication to the previous number of
the journal, in which certain phenomena of cross-breeding are
discussed in relation to the 'Origin of Species.' Mr. Westwood wrote in reply (Feb. 11), and adduced further evidence
against the doctrine of descent, such as the identity of the
figures of ostriches on the ancient " Egyptian records," with
the bird as we now know it. The correspondence is hardly
worth mentioning, except as one of the very few cases in
which my father was enticed into anything resembling a
controversy.]

* *Gardeners' Chronicle*, 1860. plete impartiality, so as not to
Referred to above, at p. 260. Sir commit Lindley.
J. D. Hooker took the line of com-

Asa Gray to J. D. Hooker.

Cambridge, Mass.,
January 5th, 1860.

My DEAR HOOKER,—Your last letter, which reached me just before Christmas, has got mislaid during the upturnings in my study which take place at that season, and has not yet been discovered. I should be very sorry to lose it, for there were in it some botanical mems. which I had not secured. . .

The principal part of your letter was high laudation of Darwin's book.

Well, the book has reached me, and I finished its careful perusal four days ago; and I freely say that your laudation is not out of place.

It is done in a *masterly manner*. It might well have taken twenty years to produce it. It is crammed full of most interesting matter—thoroughly digested—well expressed—close, cogent, and taken as a system it makes out a better case than I had supposed possible. . . .

Agassiz, when I saw him last, had read but a part of it. He says it is *poor—very poor !!* (entre nous). The fact [is] he is very much annoyed by it, ... and I do not wonder at it. To bring all *ideal* system within the domain of science, and give good physical or natural explanations of all his capital points, is as bad as to have Forbes take the glacier materials . . . and give scientific explanation of all the phenomena.

Tell Darwin all this. I will write to him when I get a chance. As I have promised, he and you shall have fair-play here. . . . I must myself write a review of Darwin's book for 'Silliman's Journal' (the more so that I suspect Agassiz means to come out upon it) for the next (March) No., and I am now setting about it (when I ought to be every moment working the Expl[oring] Expedition Compositæ, which I know far more about). And really it is no easy job as you may well imagine.

I doubt if I shall please you altogether. I know I shall not please Agassiz at all. I hear another reprint is in the Press, and the book will excite much attention here, and some controversy. . . .

C. Darwin to Asa Gray.

Down, January 28th [1860].

MY DEAR GRAY,—Hooker has forwarded to me your letter to him ; and I cannot express how deeply it has gratified me. To receive the approval of a man whom one has long sincerely respected, and whose judgment and knowledge are most universally admitted, is the highest reward an author can possibly wish for ; and I thank you heartily for your most kind expressions.

I have been absent from home for a few days, and so could not earlier answer your letter to me of the 10th of January. You have been extremely kind to take so much trouble and interest about the edition. It has been a mistake of my publisher not thinking of sending over the sheets. I had entirely and utterly forgotten your offer of receiving the sheets as printed off. But I must not blame my publisher, for had I remembered your most kind offer I feel pretty sure I should not have taken advantage of it ; for I never dreamed of my book being so successful with general readers : I believe I should have laughed at the idea of sending the sheets to America.*

After much consideration, and on the strong advice of Lyell and others, I have resolved to have the present book as it is (excepting correcting errors, or here and there inserting short

* In a letter to Mr. Murray, 1860, my father wrote :—" I am amused by Asa Gray's account of the excitement my book has made amongst naturalists in the U. States. Agassiz has denounced it in a newspaper, but yet in such terms that it is in fact a fine advertisement ! " This seems to refer to a lecture given before the Mercantile Library Association.

sentences) and to use all my strength, *which is but little*, to bring out the first part (forming a separate volume, with index, &c.) of the three volumes which will make my bigger work ; so that I am very unwilling to take up time in making corrections for an American edition. I enclose a list of a few corrections in the second reprint, which you will have received by this time complete, and I could send four or five corrections or additions of equally small importance, or rather of equal brevity. I also intend to write a *short* preface with a brief history of the subject. These I will set about, as they must some day be done, and I will send them to you in a short time —the few corrections first, and the preface afterwards, unless I hear that you have given up all idea of a separate edition. You will then be able to judge whether it is worth having the new edition with *your review prefixed*. Whatever be the nature of your review, I assure you I should feel it a *great* honour to have my book thus preceded. . . . . .

Asa Gray to C. Darwin.

Cambridge, January 23rd, 1860.

MY DEAR DARWIN,—You have my hurried letter telling you of the arrival of the remainder of the sheets of the reprint, and of the stir I had made for a reprint in Boston. Well, all looked pretty well, when, lo, we found that a second New York publishing house had announced a reprint also! I wrote then to both New York publishers, asking them to give way to the *author* and his reprint of a revised edition. I got an answer from the Harpers that they withdraw—from the Appletons that they had got the book *out* (and the next day I saw a copy) ; but that, " if the work should have any considerable sale, we certainly shall be disposed to pay the author reasonably and liberally."

The Appletons being thus out with their reprint, the Boston house declined to go on. So I wrote to the Appletons taking

them at their word, offering to aid their reprint, to give them
the use of the alterations in the London reprint, as soon as I
find out what they are, &c. &c. And I sent them the first
leaf, and asked them to insert in their future issue the additional matter from Butler,* which tells just right. So there
the matter stands. If you furnish any matter in advance of
the London third edition, I will make them pay for it.

I may get something for you. All got is clear gain ; but it
will not be very much, I suppose.

Such little notices in the papers here as have yet appeared
are quite handsome and considerate.

I hope next week to get printed sheets of my review from
New Haven, and send [them] to you, and will ask you to pass
them on to Dr. Hooker.

To fulfil your request, I ought to tell you what I think
the weakest, and what the best, part of your book. But
this is not easy, nor to be done in a word or two. The *best
part*, I think, is the *whole*, *i.e.* its *plan* and *treatment*, the vast
amount of facts and acute inferences handled as if you had a
perfect mastery of them. I do not think twenty years too
much time to produce such a book in.

Style clear and good, but now and then wants revision for
little matters (p. 97, self-fertilises *itself*, &c.).

Then your candour is worth everything to your cause. It
is refreshing to find a person with a new theory who frankly
confesses that he finds difficulties, insurmountable, at least
for the present. I know some people who never have any
difficulties to speak of.

The moment I understood your premisses, I felt sure you
had a real foundation to hold on. Well, if one admits your
premisses, I do not see how he is to stop short of your conclusions, as a probable hypothesis at least.

* A quotation from Butler's
'Analogy,' on the use of the word
natural, which in the second edition is placed with the passages
from Whewell and Bacon on p. ii,
opposite the title-page.

It naturally happens that my review of your book does not exhibit anything like the full force of the impression the book has made upon me.   Under the circumstances I suppose I do your theory more good here, by bespeaking for it a fair and favourable consideration, and by standing non-committed as to its full conclusions, than I should if I announced myself a convert ; nor could I say the latter, with truth.

Well, what seems to me the weakest point in the book is the attempt to account for the formation of organs, the making of eyes, &c., by natural selection.   Some of this reads quite Lamarckian.

The chapter on *Hybridism* is not a *weak*, but a *strong* chapter.   You have done wonders there.   But still you have not accounted, as you may be held to account, for divergence up to a certain extent producing increased fertility of the crosses, but carried one short almost imperceptible step more, giving rise to sterility, or reversing the tendency.   Very likely you are on the right track ; but you have something to do yet in that department.

Enough for the present.

. . . . . . I am not insensible to your compliments, the very high compliment which you pay me in valuing my opinion. You evidently think more of it than I do, though from the way I write [to] you, and especially [to] Hooker, this might not be inferred from the reading of my letters.

I am free to say that I never learnt so much from one book as I have from yours.   There remain a thousand things I long to say about it.

Ever yours,
ASA GRAY.

C. Darwin to Asa Gray.

[February ? 1860.]

. . . . . Now I will just run through some points in your letter.   What you say about my book gratifies me most deeply, and I wish I could feel all was deserved by me.   I quite think a review from a man, who is not an entire convert, if fair and moderately favourable, is in all respects the best kind of review.   About the weak points I agree.   The eye to this day gives me a cold shudder, but when I think of the fine known gradations, my reason tells me I ought to conquer the cold shudder.

Pray kindly remember and tell Prof. Wyman how very grateful I should be for any hints, information, or criticisms. I have the highest respect for his opinion.   I am so sorry about Dana's health.   I have already asked him to pay me a visit.

Farewell, you have laid me under a load of obligation—not that I feel it a load.   It is the highest possible gratification to me to think that you have found my book worth reading and reflection ; for you and three others I put down in my own mind as the judges whose opinions I should value most of all.

My dear Gray, yours most sincerely,

C. DARWIN.

P.S.—I feel pretty sure, from my own experience, that if you are led by your studies to keep the subject of the origin of species before your mind, you will go further and further in your belief.   It took me long years, and I assure you I am astonished at the impression my book has made on many minds.   I fear twenty years ago I should not have been half as candid and open to conviction.

C. Darwin to J. D. Hooker.

Down [January 31st, 1860].

MY DEAR HOOKER,—I have resolved to publish a little sketch of the progress of opinion on the change of species. Will you or Mrs. Hooker do me the favour to copy *one* sentence out of Naudin's paper in the 'Revue Horticole,' 1852, p. 103, namely, that on his principle of Finalité.   Can

VOL. II.   T

you let me have it soon, with those confounded dashes over
the vowels put in carefully? Asa Gray, I believe, is going to
get a second edition of my book, and I want to send this little
preface over to him soon. I did not think of the necessity of
having Naudin's sentence on finality, otherwise I would have
copied it.

Yours affectionately,

C. DARWIN.

P.S.—I shall end by just alluding to your Australian
Flora Introduction. What was the date of publication:
December 1859, or January 1860? Please answer this.

My preface will also do for the French edition, which, *I
believe*, is agreed on.

C. Darwin to J. D. Hooker.

February [1860].

. . . . As the 'Origin' now stands, Harvey's * is a good
hit against my talking so much of the insensibly fine gradations; and certainly it has astonished me that I should be
pelted with the fact, that I had not allowed abrupt and great
enough variations under nature. It would take a good deal
more evidence to make me admit that forms have often
changed by *saltum*.

* William Henry Harvey was
descended from a Quaker family of
Youghal, and was born in February, 1811, at Summerville, a
country house on the banks of the
Shannon. He died at Torquay in
1866. In 1835, Harvey went to
Africa (Table Bay) to pursue his
botanical studies, the results of
which were given in his 'Genera of
South African Plants.' In 1838,
ill-health compelled him to obtain
leave of absence, and return to
England for a time; in 1840 he
returned to Cape Town, to be again
compelled by illness to leave. In
1843 he obtained the appointment
of Botanical Professor at Trinity
College, Dublin. In 1854, 1855,
and 1856 he visited Australia, New
Zealand, the Friendly and Fiji
Islands. In 1857 Dr. Harvey
reached home, and was appointed
the successor of Professor Allman
to the Chair of Botany in Dublin
University. He was author of
several botanical works, princi-
pally on Algæ.—(From a Memoir
published in 1869.)

Have you seen Wollaston's attack in the ' Annals'? *   The stones are beginning to fly.   But Theology has more to do with these two attacks than Science. . . .

[In the above letter a paper by Harvey in the *Gardeners' Chronicle*, Feb. 18, 1860, is alluded to. He describes a case of monstrosity in *Begonia frigida*, in which the " sport " differed so much from a normal Begonia that it might have served as the type of a distinct natural order.   Harvey goes on to argue that such a case is hostile to the theory of natural selection, according to which changes are not supposed to take place *per saltum*, and adds that " a few such cases would overthrow it [Mr. Darwin's hypothesis] altogether."   In the following number of the *Gardeners' Chronicle* Sir J. D. Hooker showed that Dr. Harvey had misconceived the bearing of the Begonia case, which he further showed to be by no means calculated to shake the validity of the doctrine of modification by means of natural selection.   My father mentions the Begonia case in a letter to Lyell (Feb. 18, 1860) :—

" I send by this post an attack in the *Gardeners' Chronicle*, by Harvey (a first-rate Botanist, as you probably know).   It seems to me rather strange ; he assumes the permanence of monsters, whereas, monsters are generally sterile, and not often inheritable.   But grant his case, it comes that I have been too cautious in not admitting great and sudden varia-tions.   Here again comes in the mischief of my *abstract*.   In the fuller MS. I have discussed a parallel case of a normal fish like a monstrous gold-fish."

With reference to Sir J. D. Hooker's reply, my father wrote :]

Down [February 26th, 1860].

MY DEAR HOOKER,—Your answer to Harvey seems to me *admirably* good.   You would have made a gigantic fortune as

* ' Annals and Magazine of Natural History,' 1860.

a barrister. What an omission of Harvey's about the graduated state of the flowers! But what strikes me most is that surely I ought to know my own book best, yet, by Jove, you have brought forward ever so many arguments which I did not think of! Your reference to classification (viz. I presume to such cases as Aspicarpa) is *excellent*, for the monstrous Begonia no doubt in all details would be a Begonia. I did not think of this, nor of the *retrograde* step from separated sexes to an hermaphrodite state; nor of the lessened fertility of the monster. Proh pudor to me.

The world would say what a lawyer has been lost in a *mere* botanist!

Farewell, my dear master in my own subject,

Yours affectionately,

C. DARWIN.

I am so heartily pleased to see that you approve of the chapter on Classification.

I wonder what Harvey will say. But no one hardly, I think, is able at first to see when he is beaten in an argument.

[The following letters refer to the first translation (1860) of the 'Origin of Species' into German, which was superintended by H. G. Bronn, a good zoologist and palæontologist, who was at the time at Freiburg, but afterwards Professor at Heidelberg. I have been told that the translation was not a success, it remained an obvious translation, and was correspondingly unpleasant to read. Bronn added to the translation an appendix on the difficulties that occurred to him. For instance, how can natural selection account for differences between species, when these differences appear to be of no service to their possessors; e.g., the length of the ears and tail, or the folds in the enamel of the teeth of various species of rodents? Krause, in his book, 'Charles Darwin,' p. 91, criticises Bronn's conduct in this matter, but it will be seen that my father actually suggested the addition of Bronn's remarks. A more serious charge against Bronn made by Krause (*op. cit.* p. 87) is that he left out passages of which he did not approve, as, for instance, the passage ('Origin,' first edition, p. 488) "Light will be thrown on the origin of man and his history." I have no evidence as to whether my father did or did not know of these alterations.]

C. Darwin to H. G. Bronn.

Down, Feb. 4 [1860].

DEAR AND MUCH HONOURED SIR,—I thank you sincerely for your most kind letter; I feared that you would much disapprove of the 'Origin,' and I sent it to you merely as a mark of my sincere respect. I shall read with much interest your work on the productions of Islands whenever I receive it. I thank you cordially for the notice in the 'Neues Jahrbuch für Mineralogie,' and still more for speaking to Schweitzerbart about a translation; for I am most anxious that the great and intellectual German people should know something about my book.

I have told my publisher to send immediately a copy of the *new* * edition to Schweitzerbart, and I have written to Schweitzerbart that I give up all right to profit for myself, so that I hope a translation will appear. I fear that the book will be difficult to translate, and if you could advise Schweitzerbart about a *good* translator, it would be of very great service. Still more, if you would run your eye over the more difficult parts of the translation; but this is too great a favour to expect. I feel sure that it will be difficult to translate, from being so much condensed.

Again I thank you for your noble and generous sympathy, and I remain, with entire respect,

Yours, truly obliged,
C. DARWIN.

* Second edition.

P.S.—The new edition has some few corrections, and I will send in MS. some additional corrections, and a short historical preface, to Schweitzerbart.

How interesting you could make the work by *editing* (I do not mean translating) the work, and appending notes of *refutation* or confirmation. The book has sold so very largely in England, that an editor would, I think, make profit by the translation.

C. Darwin to H. G. Bronn.

Down, Feb. 14 [1860].

MY DEAR AND MUCH HONOURED SIR, — I thank you cordially for your extreme kindness in superintending the translation. I have mentioned this to some eminent scientific men, and they all agree that you have done a noble and generous service. If I am proved quite wrong, yet I comfort myself in thinking that my book may do some good, as truth can only be known by rising victorious from every attack. I thank you also much for the review, and for the kind manner in which you speak of me. I send with this letter some corrections and additions to M. Schweitzerbart, and a short historical preface. I am not much acquainted with German authors, as I read German very slowly ; therefore I do not know whether any Germans have advocated similar views with mine ; if they have, would you do me the favour to insert a foot-note to the preface ? M. Schweitzerbart has now the reprint ready for a translator to begin. Several scientific men have thought the term " Natural Selection " good, because its meaning is *not* obvious, and each man could not put on it his own interpretation, and because it at once connects variation under domestication and nature. Is there any analogous term used by German breeders of animals ? " Adelung," ennobling, would, perhaps, be too metaphorical. It is folly in me, but I cannot help doubting whether " Wahl der Lebensweise " expresses my notion. It leaves the impression on my mind of the Lamarckian doctrine (which I reject) of habits of life being all-important. Man has altered, and thus improved the English race-horse by *selecting* successive fleeter individuals ; and I believe, owing to the struggle for existence, that similar *slight* variations in a wild horse, *if advantageous to it*, would be *selected* or *preserved* by nature ; hence Natural Selection. But I apologise for troubling you with these remarks on the importance of choosing good German terms for " Natural Selection." With my heartfelt thanks, and with sincere respect,

I remain, dear Sir, yours very sincerely,
CHARLES DARWIN.

C. Darwin to H. G. Bronn.

Down July 14 [1860].

DEAR AND HONOURED SIR,—On my return home, after an absence of some time, I found the translation of the third part * of the 'Origin,' and I have been delighted to see a final chapter of criticisms by yourself. I have read the first few paragraphs and final paragraph, and am perfectly contented, indeed more than contented, with the generous and candid spirit with which you have considered my views. You speak with too much praise of my work. I shall, of course, carefully read the whole chapter ; but though I can read descriptive books like Gaertner's pretty easily, when any reasoning comes in, I find German excessively difficult to understand. At some *future* time I should very much like to hear how my book has been received in Germany, and I most sincerely hope M. Schweitzerbart will not lose money by the publication. Most of the reviews have been bitterly opposed to me in England, yet I have made some converts, and *several* naturalists who would not believe in a word of it, are now

---

* The German translation was published in three pamphlet-like numbers.

coming slightly round, and admit that natural selection may have done something. This gives me hope that more will ultimately come round to a certain extent to my views.

I shall ever consider myself deeply indebted to you for the immense service and honour which you have conferred on me in making the excellent translation of my book. Pray believe me, with most sincere respect,

Dear Sir, yours gratefully,

CHARLES DARWIN.

C. Darwin to C. Lyell.

Down [February 12th, 1860].

. . . I think it was a great pity that Huxley wasted so much time in the lecture on the preliminary remarks ; . . . but his lecture seemed to me very fine and very bold. I have remonstrated (and he agrees) against the impression that he would leave, that sterility was a universal and infallible criterion of species.

You will, I am sure, make a grand discussion on man. I am so glad to hear that you and Lady Lyell will come here. Pray fix your own time ; and if it did not suit us we would say so. We could then discuss man well. . . .

How much I owe to you and Hooker ! I do not suppose I should hardly ever have published had it not been for you.

[The lecture referred to in the last letter was given at the Royal Institution, February 10, 1860. The following letter was written in reply to Mr. Huxley's request for information about breeding, hybridisation, &c. It is of interest as giving a vivid retrospect of the writer's experience on the subject.]

C. Darwin to T. H. Huxley.

Ilkley, Yorks, Nov. 27 [1859].

MY DEAR HUXLEY,—Gärtner grand, Kölreuter grand, but papers scattered through many volumes and very lengthy. I

had to make an abstract of the whole. Herbert's volume on
Amaryllidaceæ very good, and two excellent papers in the
'Horticultural Journal.' For animals, no résumé to be trusted
at all; facts are to be collected from all original sources.*
I fear my MS. for the bigger book (twice or thrice as long
as in present book), with all references, would be illegible,
but it would save you infinite labour; of course I would
gladly lend it, but I have no copy, so care would have to be
taken of it. But my accursed handwriting would be fatal,
I fear.

About breeding, I know of no one book. I did not think
well of Lowe, but I can name none better. Youatt I look at
as a far better and *more practical* authority; but then his views
and facts are scattered through three or four thick volumes.
I have picked up most by reading really numberless special
treatises and *all* agricultural and horticultural journals; but
it is a work of long years. *The difficulty is to know what to
trust.* No one or two statements are worth a farthing; the
facts are so complicated. I hope and think I have been
really cautious in what I state on this subject, although all
that I have given, as yet, is *far* too briefly. I have found it
very important associating with fanciers and breeders. For
instance, I sat one evening in a gin palace in the Borough
amongst a set of pigeon fanciers, when it was hinted that
Mr. Bull had crossed his Pouters with Runts to gain size; and

* This caution is exemplified in
the following extract from an earlier
letter to Professor Huxley :—" The
inaccuracy of the blessed gang (of
which I am one) of compilers passes
all bounds. *Monsters* have frequently been described as hybrids
without a tittle of evidence. I must
give one other case to show how
we jolly fellows work. A Belgian
Baron (I forget his name at this
moment) crossed two distinct geese
and got *seven* hybrids, which he
proved subsequently to be quite
sterile; well, compiler the first,
Chevreul, says that the hybrids were
propagated for *seven* generations
inter se. Compiler second (Morton)
mistakes the French name, and
gives Latin names for two more
distinct geese, and says *Chevreul*
himself propagated them *inter se*
for seven generations; and the latter
statement is copied from book to
book."

if you had seen the solemn, the mysterious, and awful shakes
of the head which all the fanciers gave at this scandalous
proceeding, you would have recognised how little crossing
has had to do with improving breeds, and how dangerous for
endless generations the process was. All this was brought
home far more vividly than by pages of mere statements, &c.
But I am scribbling foolishly. I really do not know how to
advise about getting up facts on breeding and improving
breeds. Go to Shows is one way. Read *all* treatises on any
one domestic animal, and believe nothing without largely
confirmed. For your lectures I can give you a few amusing
anecdotes and sentences, if you want to make the audience
laugh.

I thank you particularly for telling me what naturalists
think. If we can once make a compact set of believers we
shall in time conquer. I am *eminently* glad Ramsay is on
our side, for he is, in my opinion, a first-rate geologist. I sent
him a copy. I hope he got it. I shall be very curious to
hear whether any effect has been produced on Prestwich; I
sent him a copy, not as a friend, but owing to a sentence
or two in some paper, which made me suspect he was
doubting.

Rev. C. Kingsley has a mind to come round. Quatrefages
writes that he goes some long way with me ; says he exhibited
diagrams like mine. With most hearty thanks,

Yours very tired,

C. DARWIN.

[I give the conclusion of Professor Huxley's lecture, as
being one of the earliest, as well as one of the most eloquent,
of his utterances in support of the 'Origin of Species':

"I have said that the man of science is the sworn interpreter of nature in the high court of reason. But of what
avail is his honest speech, if ignorance is the assessor of the
judge, and prejudice the foreman of the jury ? I hardly know

of a great physical truth, whose universal reception has not
been preceded by an epoch in which most estimable persons have maintained that the phenomena investigated were
directly dependent on the Divine Will, and that the attempt
to investigate them was not only futile, but blasphemous.
And there is a wonderful tenacity of life about this sort of
opposition to physical science. Crushed and maimed in every
battle, it yet seems never to be slain ; and after a hundred
defeats it is at this day as rampant, though happily not so
mischievous, as in the time of Galileo.

"But to those whose life is spent, to use Newton's noble
words, in picking up here a pebble and there a pebble on the
shores of the great ocean of truth—who watch, day by day,
the slow but sure advance of that mighty tide, bearing on its
bosom the thousand treasures wherewith man ennobles and
beautifies his life—it would be laughable, if it were not so sad,
to see the little Canutes of the hour enthroned in solemn
state, bidding that great wave to stay, and threatening to
check its beneficent progress. The wave rises and they fly ;
but, unlike the brave old Dane, they learn no lesson of
humility : the throne is pitched at what seems a safe distance,
and the folly is repeated.

"Surely it is the duty of the public to discourage anything
of this kind, to discredit these foolish meddlers who think
they do the Almighty a service by preventing a thorough
study of His works.

"The Origin of Species is not the first, and it will not be
the last, of the great questions born of science, which will
demand settlement from this generation. The general mind
is seething strangely, and to those who watch the signs of the
times, it seems plain that this nineteenth century will see
revolutions of thought and practice as great as those which
the sixteenth witnessed. Through what trials and sore con-
tests the civilised world will have to pass in the course of this
new reformation, who can tell ?

"But I verily believe that come what will, the part which England may play in the battle is a grand and a noble one. She may prove to the world that, for one people, at any rate, despotism and demagogy are not the necessary alternatives of government; that freedom and order are not incompatible; that reverence is the handmaid of knowledge; that free discussion is the life of truth, and of true unity in a nation.

"Will England play this part? That depends upon how you, the public, deal with science. Cherish her, venerate her, follow her methods faithfully and implicitly in their application to all branches of human thought, and the future of this people will be greater than the past.

"Listen to those who would silence and crush her, and I fear our children will see the glory of England vanishing like Arthur in the mist; they will cry too late the woful cry of Guinever:—

> ' It was my duty to have loved the highest;
> It surely was my profit had I known;
> It would have been my pleasure had I seen.' "]

C. Darwin to C. Lyell.

Down [February 15th, 1860].

. . . I am perfectly convinced (having read this morning) that the review in the 'Annals'* is by Wollaston; no one else in the world would have used so many parentheses. I

* Annals and Mag. of Nat. Hist. third series, vol. 5, p. 132. My father has obviously taken the expression " pestilent " from the following passage (p. 138) : " But who is this Nature, we have a right to ask, who has such tremendous power, and to whose efficiency such marvellous performances are ascribed? What are her image and attributes, when dragged from her wordy lurking-place? Is she ought but a pestilent abstraction, like dust cast in our eyes to obscure the workings of an Intelligent First Cause of all?" The reviewer pays a tribute to my father's candour, " so manly and outspoken as almost to 'cover a multitude of sins.'" The parentheses (to which allusion is made above) are so frequent as to give a characteristic appearance to Mr. Wollaston's pages.

have written to him, and told him that the "pestilent" fellow thanks him for his kind manner of speaking about him. I have also told him that he would be pleased to hear that the Bishop of Oxford says it is the most unphilosophical * work he ever read. The review seems to me clever, and only misinterprets me in a few places. Like all hostile men, he passes over the explanation given of Classification, Morphology, Embryology, and Rudimentary Organs, &c. I read Wallace's paper in MS.,† and thought it admirably good; he does not know that he has been anticipated about the depth of intervening sea determining distribution. . . . The most curious point in the paper seems to me that about the African character of the Celebes productions, but I should require further confirmation. . . .

Henslow is staying here; I have had some talk with him; he is in much the same state as Bunbury,‡ and will go a very little way with us, but brings up no real argument against going further. He also shudders at the eye! It is really curious (and perhaps is an argument in our favour) how differently different opposers view the subject. Henslow used to rest his opposition on the imperfection of the Geological Record, but he now thinks nothing of this, and says I have got well out of it ; I wish I could quite agree with him. Baden Powell says he never read anything so conclusive as my statement about the eye!! A stranger writes to me about sexual selection, and regrets that I boggle about such a trifle as the brush of hair on the male turkey, and so on. As L. Jenyns has a really philosophical mind, and as you say you like to see everything, I send an old letter of his. In a later letter to Henslow, which I have seen, he is more candid than any opposer I have heard of, for he says, though he *cannot* go so

* Another version of the words is given by Lyell, to whom they were spoken, viz. "the most illogical book ever written."—'Life,' vol. ii. p. 358.

† "On the Zoological Geography of the Malay Archipelago."—Linn. Soc. Journ. 1860.

‡ The late Sir Charles Bunbury, well known as a Palæo-botanist.

far as I do, yet he can give no good reason why he should not. It is funny how each man draws his own imaginary line at which to halt. It reminds me so vividly what I was told * about you when I first commenced geology—to believe a *little*, but on no account to believe all.

Ever yours affectionately,

C. DARWIN.

C. Darwin to Asa Gray.

Down, February 18th [1860].

MY DEAR GRAY,—I received about a week ago two sheets of your Review ;† read them, and sent them to Hooker ; they are now returned and re-read with care, and to-morrow I send them to Lyell. Your Review seems to me *admirable ;* by far the best which I have read. I thank you from my heart both for myself, but far more for the subject's sake. Your contrast between the views of Agassiz and such as mine is very curious and instructive.‡ By the way, if Agassiz writes anything on the subject, I hope you will tell me. I am charmed with your metaphor of the streamlet never running against the force of gravitation. Your distinction between an hypothesis and theory seems to me very ingenious ; but I do not think it is ever followed. Every one now speaks of the undulatory *theory* of light ; yet the ether is itself hypothetical, and the undulations are inferred only from explaining the phenomena of light. Even in the *theory* of gravitation is the attractive power in any way known, except by explaining the fall of the apple, and the movements of the Planets? It seems to me that an hypothesis is *developed* into a theory solely by explaining an ample lot of facts. Again and again I

* By Professor Henslow.

† The 'American Journal of Science and Arts,' March 1860. Reprinted in 'Darwiniana,' 1876.

‡ The contrast is briefly summed up thus : "The theory of Agassiz regards the origin of species and their present general distribution over the world as equally primordial, equally supernatural ; that of Darwin as equally derivative, equally natural."—'Darwiniana,' p. 14.

thank you for your generous aid in discussing a view, about which you very properly hold yourself unbiassed.

My dear Gray, yours most sincerely,

C. DARWIN.

P.S.—Several clergymen go far with me. Rev. L. Jenyns, a very good naturalist. Henslow will go a very little way with me, and is not shocked at me. He has just been visiting me.

[With regard to the attitude of the more liberal representatives of the Church, the following letter (already referred to) from Charles Kingsley is of interest :]

C. Kingsley to C. Darwin.

Eversley Rectory, Winchfield,
November 18th, 1859.

DEAR SIR,—I have to thank you for the unexpected honour of your book. That the Naturalist whom, of all naturalists living, I most wish to know and to learn from, should have sent a scientist like me his book, encourages me at least to observe more carefully, and think more slowly.

I am so poorly (in brain), that I fear I cannot read your book just now as I ought. All I have seen of it *awes* me ; both with the heap of facts and the prestige of your name, and also with the clear intuition, that if you be right, I must give up much that I have believed and written.

In that I care little. Let God be true, and every man a liar ! Let us know what *is*, and, as old Socrates has it, ἕπεσθαι τῷ λόγῳ—follow up the villainous shifty fox of an argument, into whatsoever unexpected bogs and brakes he may lead us, if we do but run into him at last.

From two common superstitions, at least, I shall be free while judging of your book :—

(1.) I have long since, from watching the crossing of domesticated animals and plants, learnt to disbelieve the dogma of the permanence of species.

(2.) I have gradually learnt to see that it is just as noble a conception of Deity, to believe that He created primal forms capable of self development into all forms needful *pro tempore* and *pro loco*, as to believe that He required a fresh act of intervention to supply the *lacunas* which He Himself had made. I question whether the former be not the loftier thought.

Be it as it may, I shall prize your book, both for itself, and as a proof that you are aware of the existence of such a person as

Your faithful servant,

C. KINGSLEY.

[My father's old friend, the Rev. J. Brodie Innes, of Milton Brodie, who was for many years Vicar of Down, writes in the same spirit:

"We never attacked each other.   Before I knew Mr. Darwin I had adopted, and publicly expressed, the principle that the study of natural history, geology, and science in general, should be pursued without reference to the Bible.   That the Book of Nature and Scripture came from the same Divine source, ran in parallel lines, and when properly understood would never cross. . . . . .

"His views on this subject were very much to the same effect from his side.   Of course any conversations we may have had on purely religious subjects are as sacredly private now as in his life ; but the quaint conclusion of one may be given.   We had been speaking of the apparent contradiction of some supposed discoveries with the Book of Genesis ; he said, 'you are (it would have been more correct to say you ought to be) a theologian, I am a naturalist, the lines are separate.   I endeavour to discover facts without considering what is said in the Book of Genesis.   I do not attack Moses, and I think Moses can take care of himself.'   To the same effect he wrote more recently, 'I cannot remember that I ever published a word directly against religion or the clergy ; but if you were to read a little pamphlet which I received a couple of days ago by a clergyman, you would laugh, and admit that I had some excuse for bitterness.   After abusing me for two or three pages, in language sufficiently plain and emphatic to have satisfied any reasonable man, he sums up by saying that he has vainly searched the English language to find terms to express his contempt for me and all Darwinians.'   In another letter, after I had left Down, he writes, ' We often differed, but you are one of those rare mortals from whom one can differ and yet feel no shade of animosity, and that is a thing [of] which I should feel very proud, if any one could say [it] of me.'

"On my last visit to Down, Mr. Darwin said, at his dinner-table, 'Brodie Innes and I have been fast friends for thirty years, and we never thoroughly agreed on any subject but once, and then we stared hard at each other, and thought one of us must be very ill.'"]

C. Darwin to C. Lyell.

Down, February 23rd [1860].

MY DEAR LYELL,—That is a splendid answer of the father of Judge Crampton.   How curious that the Judge should have hit on exactly the same points as yourself.   It shows me what a capital lawyer you would have made, how many unjust acts you would have made appear just !   But how much grander a field has science been than the law, though the latter might have made you Lord Kinnordy.   I will, if there be another edition, enlarge on gradation in the eye, and on all forms coming from one prototype, so as to try and make both less glaringly improbable. . . .

With respect to Bronn's objection that it cannot be shown how life arises, and likewise to a certain extent Asa Gray's remark that natural selection is not a *vera causa*, I was much

interested by finding accidentally in Brewster's 'Life of
Newton,' that Leibnitz objected to the law of gravity because
Newton could not show what gravity itself is.   As it has
chanced, I have used in letters this very same argument,
little knowing that any one had really thus objected to the law
of gravity.   Newton answers by saying that it is philosophy
to make out the movements of a clock, though you do not
know why the weight descends to the ground.   Leibnitz further objected that the law of gravity was opposed to Natural
Religion !   Is this not curious ?   I really think I shall use the
facts for some introductory remarks for my bigger book.

. . . You ask (I see) why we do not have monstrosities in
higher animals ; but when they live they are almost always
sterile (even giants and dwarfs are *generally* sterile), and we do
not know that Harvey's monster would have bred.   There is
I believe only one case on record of a peloric flower being
fertile, and I cannot remember whether this reproduced itself.

To recur to the eye.   I really think it would have been dishonest, not to have faced the difficulty ; and worse (as Talley-
rand would have said), it would have been impolitic I think,
for it would have been thrown in my teeth, as H. Holland
threw the bones of the ear, till Huxley shut him up by showing
what a fine gradation occurred amongst living creatures.

I thank you much for your most pleasant letter.

Yours affectionately,

C. DARWIN.

P.S.—I send a letter by Herbert Spencer, which you can
read or not as you think fit.   He puts, to my mind, the
philosophy of the argument better than almost any one,
at the close of the letter.   I could make nothing of Dana's
idealistic notions about species ; but then, as Wollaston says,
I have not a metaphysical head.

By the way, I have thrown at Wollaston's head, a paper by
Alexander Jordan, who demonstrates metaphysically that all
our cultivated races are God-created species.

Wollaston misrepresents accidentally, to a wonderful extent, some passages in my book. He reviewed, without relooking at certain passages.

C. Darwin to C. Lyell.

.Down, February 25th [1860].

. . . . . I cannot help wondering at your zeal about my book. I declare to heaven you seem to care as much about my book as I do myself. You have no right to be so eminently unselfish! I have taken off my spit [*i.e.* file] a letter of Ramsay's, as every geologist convert I think very important. By the way, I saw some time ago a letter from H. D. Rogers* to Huxley, in which he goes very far with us. . . . .

C. Darwin to J. D. Hooker.

Down, Saturday March 3rd, [1860].

MY DEAR HOOKER,—What a day's work you had on that Thursday! I was not able to go to London till Monday, and then I was a fool for going, for, on Tuesday night, I had an attack of fever (with a touch of pleurisy), which came on like a lion, but went off as a lamb, but has shattered me a good bit.

I was much interested by your last note. . . . I think you expect too much in regard to change of opinion on the subject of Species. One large class of men, more especially I suspect of naturalists, never will care about *any* general question, of which old Gray, of the British Museum, may be taken as a type; and secondly, nearly all men past a moderate age, either in actual years or in mind, are, I am fully convinced, incapable of looking at facts under a new point of view. Seriously, I am astonished and rejoiced at the progress which

* Professor of Geology in the University of Glasgow. Born in the United States 1809, died 1866.

the subject has made ; look at the enclosed memorandum.*
—— says my book will be forgotten in ten years, perhaps so ;
but, with such a list, I feel convinced the subject will not.
The outsiders, as you say, are strong.

You say that you think that Bentham is touched, "but,
like a wise man, holds his tongue." Perhaps you only mean
that he cannot decide, otherwise I should think such silence
the reverse of magnanimity ; for if others behaved the same
way, how would opinion ever progress? It is a dereliction of
actual duty.†

I am so glad to hear about Thwaites.‡ . . . I have had an
astounding letter from Dr. Boott ;§ it might be turned into
ridicule against him and me, so I will not send it to any one.
He writes in a noble spirit of love of truth.

I wonder what Lindley thinks ; probably too busy to read
or think on the question.

I am vexed about Bentham's reticence, for it would have
been of real value to know what parts appeared weakest to a
man of his powers of observation.

Farewell, my dear Hooker, yours affectionately,

C. DARWIN.

P.S.—Is not Harvey in the class of men who do not at all
care for generalities? I remember your saying you could

* See table of names, p. 293.

† In a subsequent letter to Sir
J. D. Hooker (March 12th, 1860),
my father wrote, "I now quite understand Bentham's silence."

‡ Dr. G. J. K. Thwaites, who
was born in 1811, established a
reputation in this country as an
expert microscopist and an acute
observer, working especially at
cryptogamic botany. On his appointment as Director of the
Botanic Gardens at Peradenyia,
Ceylon, Dr. Thwaites devoted himself to the flora of Ceylon. As a
result of this he has left numerous
and valuable collections, a description of which he embodied in his
' Enumeratio Plantarum Zeylaniae '
(1864). Dr. Thwaites was a Fellow
of the Linnean Society, but beyond
the above facts, little seems to have
been recorded of his life. His death
occurred in Ceylon on September
11th, 1882, in his seventy-second
year. *Athenæum*, October 14th,
1882, p. 500.

§ The letter is enthusiastically
laudatory, and obviously full of
genuine feeling.

not get him to write on Distribution. I have found his
works very unfruitful in every respect.

[Here follows the memorandum referred to :]

Geologists.	Zoologists and Palæontologists.	Physiologists.	Botanists.
Lyell.	Huxley.	Carpenter.	Hooker.
Ramsay.*	J. Lubbock.	Sir H. Holland (to large extent).	H. C. Watson.
Jukes.†	L. Jenyns (to large extent).		Asa Gray (to some extent).
H. D. Rogers.	Searles Wood.‡		Dr. Boott (to large extent).
			Thwaites.

[The following letter is of interest in connection with the
mention of Mr. Bentham in the last letter :]

G. Bentham to Francis Darwin.

25 Wilton Place, S.W.,
May 30th, 1882.

MY DEAR SIR.—In compliance with your note which I
received last night, I send herewith the letters I have from
your father. I should have done so on seeing the general
request published in the papers, but that I did not think
there were any among them which could be of any use to
you. Highly flattered as I was by the kind and friendly
notice with which Mr. Darwin occasionally honoured me, I

* Andrew Ramsay, late Director-General of the Geological Survey.

† Joseph Beete Jukes, M.A., F.R.S., born 1811, died 1869. He was educated at Cambridge, and from 1842 to 1846 he acted as naturalist to H.M.S. *Fly*, on an exploring expedition in Australia and New Guinea. He was after-wards appointed Director of the Geological Survey of Ireland. He was the author of many papers, and of more than one good hand-book of geology.

‡ Searles Valentine Wood, born Feb. 14, 1798, died 1880. Chiefly known for his work on the Mollusca of the ' Crag.'

was never admitted into his intimacy, and he therefore never made any communications to me in relation to his views and labours. I have been throughout one of his most sincere admirers, and fully adopted his theories and conclusions, notwithstanding the severe pain and disappointment they at first occasioned me. On the day that his celebrated paper was read at the Linnean Society, July 1st, 1858, a long paper of mine had been set down for reading, in which, in commenting on the British Flora, I had collected a number of observations and facts illustrating what I then believed to be a fixity in species, however difficult it might be to assign their limits, and showing a tendency of abnormal forms produced by cultivation or otherwise, to withdraw within those original limits when left to themselves. Most fortunately my paper had to give way to Mr. Darwin's, and when once that was read, I felt bound to defer mine for reconsideration; I began to entertain doubts on the subject, and on the appearance of the 'Origin of Species,' I was forced, however reluctantly, to give up my long-cherished convictions, the results of much labour and study, and I cancelled all that part of my paper which urged original fixity, and published only portions of the remainder in another form, chiefly in the 'Natural History Review.' I have since acknowledged on various occasions my full adoption of Mr. Darwin's views, and chiefly in my Presidential Address of 1863, and in my thirteenth and last address, issued in the form of a report to the British Association at its meeting at Belfast in 1874.

I prize so highly the letters that I have of Mr. Darwin's, that I should feel obliged by your returning them to me when you have done with them. Unfortunately I have not kept the envelopes, and Mr. Darwin usually only dated them by the month not by the year, so that they are not in any chronological order.

Yours very sincerely,
GEORGE BENTHAM.

C. Darwin to C. Lyell.

Down [March] 12th [1860].

MY DEAR LYELL,—Thinking over what we talked about, the high state of intellectual development of the old Grecians with the little or no subsequent improvement, being an apparent difficulty, it has just occurred to me that in fact the case harmonises perfectly with our views. The case would be a decided difficulty on the Lamarckian or Vestigian doctrine of necessary progression, but on the view which I hold of progression depending on the conditions, it is no objection at all, and harmonises with the other facts of progression in the corporeal structure of other animals. For in a state of anarchy, or despotism, or bad government, or after irruption of barbarians, force, strength, or ferocity, and not intellect, would be apt to gain the day.

We have so enjoyed your and Lady Lyell's visit.

Good-night.

C. DARWIN.

P.S.—By an odd chance (for I had not alluded even to the subject) the ladies attacked me this evening, and threw the high state of old Grecians into my teeth, as an unanswerable difficulty, but by good chance I had my answer all pat, and silenced them. Hence I have thought it worth scribbling to you. . . .

*C. Darwin to J. Prestwich.**

Down, March 12th [1860].

. . . At some future time, when you have a little leisure, and when you have read my 'Origin of Species,' I should esteem it a *singular* favour if you would send me any general criticisms. I do not mean of unreasonable length, but such

---

* Now Professor of Geology in the University of Oxford.

as you could include in a letter. I have always admired your various memoirs so much that I should be eminently glad to receive your opinion, which might be of real service to me.

Pray do not suppose that I expect to *convert* or *pervert* you; if I could stagger you in ever so slight a degree I should be satisfied; nor fear to annoy me by severe criticisms, for I have had some hearty kicks from some of my best friends. If it would not be disagreeable to you to send me your opinion, I certainly should be truly obliged. . . .

C. Darwin to Asa Gray.

Down, April 3 [1860].

. . . . I remember well the time when the thought of the eye made me cold all over, but I have got over this stage of the complaint, and now small trifling particulars of structure often make me very uncomfortable. The sight of a feather in a peacock's tail, whenever I gaze at it, makes me sick! . . .

You may like to hear about reviews on my book. Sedgwick (as I and Lyell feel *certain* from internal evidence) has reviewed me savagely and unfairly in the *Spectator*.* The notice includes much abuse, and is hardly fair in several respects. He would actually lead any one, who was ignorant of geology, to suppose that I had invented the great gaps between successive geological formations, instead of its being an almost universally admitted dogma. But my dear old friend Sedgwick, with his noble heart, is old, and is rabid with indignation. It is hard to please every one; you may remember that in my last letter I asked you to leave out about the Weald denudation : I told Jukes this (who is head man of the Irish geological survey), and he blamed me much, for he believed every word of it, and thought it not at all exaggerated! In fact, geologists have no means of gauging the infinitude of past time. There has been one prodigy of a

* See the quotations which follow the present letter.

review, namely, an *opposed* one (by Pictet,* the palæontologist,
in the Bib. Universelle of Geneva) which is *perfectly* fair and
just, and I agree to every word he says; our only difference
being that he attaches less weight to arguments in favour,
and more to arguments opposed, than I do. Of all the opposed reviews, I think this the only quite fair one, and I never
expected to see one. Please observe that I do not class your
review by any means as opposed, though you think so yourself! It has done me *much* too good service ever to appear
in that rank in my eyes. But I fear I shall weary you with
so much about my book. I should rather think there was a
good chance of my becoming the most egotistical man in all
Europe! What a proud pre-eminence! Well, you have
helped to make me so, and therefore you must forgive me if
you can.

My dear Gray, ever yours most gratefully,

C. DARWIN.

[In a letter to Sir Charles Lyell reference iṣ made to
Sedgwick's review in the *Spectator*, March 24:

"I now feel certain that Sedgwick is the author of the
article in the *Spectator*. No one else could use such abusive
terms. And what a misrepresentation of my notions! Any
ignoramus would suppose that I had *first* broached the

* François Jules Pictet, in the 'Archives des Sciences de la Bib-liothèque Universelle,' Mars 1860. The article is written in a courteous and considerate tone, and con-cludes by saying that the 'Origin' will be of real value to naturalists, especially if they are not led away by its seductive arguments to be-lieve in the dangerous doctrine of modification. A passage which seems to have struck my father as being valuable, and opposite which he has made double pencil marks and written the word "good," is worth quoting: "La théorie de M. Darwin s'accorde mal avec l'histoire des types à formes bien tranchées et définies qui paraissent n'avoir vécu que pendant un temps limité. On en pourrait citer des centaines d'exemples, tel que les reptiles volants, les ichthyosaures, les bélemnites, les ammonites, &c." Pictet was born in 1809, died 1872; he was Professor of Anatomy and Zoology at Geneva.

doctrine, that the breaks between successive formations
marked long intervals of time. It is very unfair. But poor
·dear old Sedgwick seems rabid on the question. " Demoralised understanding ! " If ever I talk with him I will tell
him that I never could believe that an inquisitor could be a
good man ; but now I know that a man may roast another,
and yet have as kind and noble a heart as Sedgwick's."

The following passages are taken from the review :

" I need hardly go on any further with these objections.
But I cannot conclude without expressing my detestation of
the theory, because of its unflinching materialism ;—because
it has deserted the inductive track, the only track that leads
to physical truth ;—because it utterly repudiates final causes,
and thereby indicates a demoralised understanding on the
part of its advocates."

" Not that I believe that Darwin is an atheist ; though I
·cannot but regard his materialism as atheistical. I think it
untrue, because opposed to the obvious course of nature, and
the very opposite of inductive truth. And I think it intensely
mischievous."

" Each series of facts is laced together by a series of
assumptions, and repetitions of the one false principle.
You cannot make a good rope out of a string of air
bubbles."

" But any startling and (supposed) novel paradox, maintained very boldly and with something of imposing plausi-
bility, produces in some minds a kind of pleasing excitement
which predisposes them in its favour ; and if they are unused
to careful reflection, and averse to the labour of accurate
investigation, they will be likely to conclude that what is
(apparently) *original*, must be a production of original *genius*,
and that anything very much opposed to prevailing notions
must be a grand *discovery*,— in short, that whatever comes
from the 'bottom of a well' must be the 'truth' supposed to
be hidden there."

In a review in the December number of 'Macmillan's Magazine,' 1860, Fawcett vigorously defended my father from the charge of employing a false method of reasoning ; a charge which occurs in Sedgwick's review, and was made at the time *ad nauseam*, in such phrases as : "This is not the true Baconian method." Fawcett repeated his defence at the meeting of the British Association in 1861.*]

C. Darwin to W. B. Carpenter.

Down, April 6th [1860].

MY DEAR CARPENTER,—I have this minute finished your review in the 'Med. Chirurg. Review.'† You must let me express my admiration at this most able essay, and I hope to God it will be largely read, for it must produce a great effect. I ought not, however, to express such warm admiration, for you give my book, I fear, far too much praise. But you have gratified me extremely ; and though I hope I do not care very much for the approbation of the non-scientific readers, I cannot say that this is at all so with respect to such few men as yourself. I have not a criticism to make, for I object to not a word ; and I admire all, so that I cannot pick out one part as better than the rest. It is all so well balanced. But it is impossible not to be struck with your extent of knowledge in geology, botany, and zoology. The extracts which you give from Hooker seem to me *excellently* chosen, and most forcible. I am so much pleased in what you say also about Lyell. In fact I am in a fit of enthusiasm, and had better write no more. With cordial thanks,

Yours very sincerely,

C. DARWIN.

---

* See an interesting letter from my father in Mr. Stephen's 'Life of   Henry Fawcett,' 1886, p. 101.
† April 1860.

C. Darwin to C. Lyell.

Down, April 10th [1860].

MY DEAR LYELL,—Thank you much for your note of the 4th ; I am very glad to hear that you are at Torquay. I should have amused myself earlier by writing to you, but I have had Hooker and Huxley staying here, and they have fully occupied my time, as a little of anything is a full dose for me. . . . There has been a plethora of reviews, and I am really quite sick of myself. There is a very long review by Carpenter in the 'Medical and Chirurg. Review,' very good and well balanced, but not brilliant. He discusses Hooker's books at as great length as mine, and makes excellent extracts ; but I could not get Hooker to feel the least interest in being praised.

Carpenter speaks of you in thoroughly proper terms. There is a *brilliant* review by Huxley, * with capital hits, but I do not know that he much advances the subject. I *think* I have convinced him that he has hardly allowed weight enough to the case of varieties of plants being in some degrees sterile.

To diverge from reviews : Asa Gray sends me from Wyman (who will write), a good case of all the pigs being black in the Everglades of Virginia. On asking about the cause, it seems (I have got capital analogous cases) that when the *black* pigs eat a certain nut their bones become red, and they suffer to a certain extent, but that the *white* pigs lose their hoofs and perish, " and we aid by *selection,* for we kill most of the young white pigs." This was said by men who could hardly read. By the way, it is a great blow to me that you cannot admit the potency of natural selection. The more I think of it, the less I doubt its power for great and small changes. I have just read the 'Edinburgh,' † which without doubt is by —— It is extremely malignant, clever, and I fear will be very damaging. He is atrociously severe on Huxley's lecture,

* 'Westminster Review,' April 1860.
† 'Edinburgh Review,' April 1860.

and very bitter against Hooker. So we three *enjoyed* it together. Not that I really enjoyed it, for it made me uncomfortable for one night; but I have got quite over it to-day. It requires much study to appreciate all the bitter spite of many of the remarks against me; indeed I did not discover all myself. It scandalously misrepresents many parts. He misquotes some passages, altering words within inverted commas. . . .

It is painful to be hated in the intense degree with which —— hates me.

Now for a curious thing about my book, and then I have done. In last Saturday's *Gardeners' Chronicle*,* a Mr. Patrick Matthew publishes a long extract from his work on 'Naval Timber and Arboriculture,' published in 1831, in which he briefly but completely anticipates the theory of Natural Selection. I have ordered the book, as some few passages are rather obscure, but it is certainly, I think, a complete but not developed anticipation! Erasmus always said that surely this would be shown to be the case some day. Anyhow, one may be excused in not having discovered the fact in a work on Naval Timber.

I heartily hope that your Torquay work may be successful. Give my kindest remembrances to Falconer, and I hope he is pretty well. Hooker and Huxley (with Mrs. Huxley) were extremely pleasant. But poor dear Hooker is tired to death of my book, and it is a marvel and a prodigy if you are not worse tired—if that be possible. Farewell, my dear Lyell,

Yours affectionately,

C. DARWIN.

C. Darwin to J. D. Hooker.

Down [April 13th, 1860].

MY DEAR HOOKER,—Questions of priority so often lead to odious quarrels, that I should esteem it a great favour if you

* April 7th, 1860.

would read the enclosed.* If you think it proper that I
should send it (and of this there can hardly be any question),
and if you think it full and ample enough, please alter the
date to the day on which you post it, and let that be soon.
The case in the *Gardeners' Chronicle* seems a *little* stronger
than in Mr. Matthew's book, for the passages are therein
scattered in three places ; but it would be mere hair-splitting
to notice that. If you object to my letter, please return it ;
but I do not expect that you will, but I thought that you
would not object to run your eye over it. My dear Hooker,
it is a great thing for me to have so good, true, and old a
friend as you. I owe much for science to my friends.

Many thanks for Huxley's lecture. The latter part seemed
to be grandly eloquent.

. . . I have gone over [the 'Edinburgh'] review again, and
compared passages, and I am astonished at the misrepre-
sentations. But I am glad I resolved not to answer. Perhaps
it is selfish, but to answer and think more on the subject is
too unpleasant. I am so sorry that Huxley by my means
has been thus atrociously attacked. I do not suppose you
much care about the gratuitous attack on you.

---

* My father wrote (*Gardeners'
Chronicle*, 1860, p. 362, April 21st) :
" I have been much interested by
Mr. Patrick Matthew's communication in the number of your paper
dated April 7th. I freely acknow-
ledge that Mr. Matthew has anticipated by many years the explanation which I have offered of
the origin of species, under the
name of natural selection. I think
that no one will feel surprised
that neither I, nor apparently any
other naturalist, had heard of Mr.
Matthew's views, considering how
briefly they are given, and that
they appeared in the appendix to
a work on Naval Timber and
Arboriculture. I can do no more
than offer my apologies to Mr.
Matthew for my entire ignorance
of his publication. If another edition of my work is called for, I will
insert to the foregoing effect." In
spite of my father's recognition of
his claims, Mr. Matthew remained
unsatisfied, and complained that
an article in the ' Saturday Analyst
and Leader' was "scarcely fair in
alluding to Mr. Darwin as the
parent of the origin of species,
seeing that I published the whole
that Mr. Darwin attempts to prove,
more than twenty-nine years ago."
—*Saturday Analyst and Leader*,
Nov. 24, 1860.

Lyell in his letter remarked that you seemed to him as if you were overworked. Do, pray, be cautious, and remember how many and many a man has done this—who thought it absurd till too late. I have often thought the same. You know that you were bad enough before your Indian journey.

C. Darwin to C. Lyell.

Down, April [1860].

MY DEAR LYELL,—I was very glad to get your nice long letter from Torquay. A press of letters prevented me writing to Wells. I was particularly glad to hear what you thought about not noticing [the 'Edinburgh'] review. Hooker and Huxley thought it a sort of duty to point out the alteration of quoted citations, and there is truth in this remark; but I so hated the thought that I resolved not to do so. I shall come up to London on Saturday the 14th, for Sir B. Brodie's party, as I have an accumulation of things to do in London, and will (if I do not hear to the contrary) call about a quarter before ten on Sunday morning, and sit with you at breakfast, but will not sit long, and so take up much of your time. I must say one more word about our quasi-theological controversy about natural selection, and let me have your opinion when we meet in London. Do you consider that the successive variations in the size of the crop of the Pouter Pigeon, which man has accumulated to please his caprice, have been due to "the creative and sustaining powers of Brahma?" In the sense that an omnipotent and omniscient Deity must order and know everything, this must be admitted; yet, in honest truth, I can hardly admit it. It seems preposterous that a maker of a universe should care about the crop of a pigeon solely to please man's silly fancies. But if you agree with me in thinking such an interposition of the Deity uncalled for, I can see no reason whatever for believing in such interpositions in the case of natural beings, in which strange and admirable peculiarities have been naturally selected for the creature's own benefit. Imagine a Pouter in a state of nature wading into the water and then, being buoyed up by its inflated crop, sailing about in search of food. What admiration this would have excited —adaptation to the laws of hydrostatic pressure, &c. &c. For the life of me I cannot see any difficulty in natural selection producing the most exquisite structure, *if such structure can be arrived at by gradation*, and I know from experience how hard it is to name any structure towards which at least some gradations are not known.

Ever yours,
C. DARWIN.

P.S.—The conclusion at which I have come, as I have told Asa Gray, is that such a question, as is touched on in this note, is beyond the human intellect, like " predestination and free will," or the " origin of evil."

C. Darwin to J. D. Hooker.

Down [April 18th, 1860].

MY DEAR HOOKER,—I return ——'s letter. . . . Some of my relations say it cannot *possibly* be ——'s article,* because the reviewer speaks so very highly of ——. Poor dear simple folk! My clever neighbour, Mr. Norman, says the article is so badly written, with no definite object, that no one will read it. . . . Asa Gray has sent me an article † from the United States, clever, and dead against me. But one argument is funny. The reviewer says, that if the doctrine were true, geological strata would be full of monsters which have failed. A very clear view this writer had of the struggle for existence !

---

* The ' Edinburgh Review.'

† ' North American Review,' April 1860. " By Professor Bowen," is written on my father's copy. The passage referred to occurs at p. 488, where the author says that we ought to find " an infinite number of other varieties—gross, rude, and purposeless— the unmeaning creations of an unconscious cause."

. . . . I am glad you like Adam Bede so much. I was charmed with it. . . .

We think you must by mistake have taken with your own numbers of the 'National Review' my precious number.* I wish you would look.

C. Darwin to Asa Gray.

Down, April 25th [1860].

MY DEAR GRAY,—I have no doubt I have to thank you for the copy of a review on the 'Origin' in the 'North American Review.' It seems to me clever, and I do not doubt will damage my book. I had meant to have made some remarks on it; but Lyell wished much to keep it, and my head is quite confused between the many reviews which I have lately read. I am sure the reviewer is wrong about bees' cells, *i.e.* about the distance; any lesser distance would do, or even greater distance, but then some of the places would lie outside the generative spheres; but this would not add much difficulty to the work. The reviewer takes a strange view of instinct: he seems to regard intelligence as a developed instinct; which I believe to be wholly false. I suspect he has never much attended to instinct and the minds of animals, except perhaps by reading.

My chief object is to ask you if you could procure for me a copy of the *New York Times* for Wednesday, March 28th. It contains *a very striking* review of my book, which I should much like to keep. How curious that the two most striking reviews (*i.e.* yours and this) should have appeared in America. This review is not really useful, but somehow is impressive. There was a good review in the 'Revue des Deux Mondes,' April 1st, by M. Laugel, said to be a very clever man.

* This no doubt refers to the January number, containing Dr. Carpenter's review of the 'Origin.'

Hooker, about a fortnight ago, stayed here a few days, and was very pleasant ; but I think he overworks himself. What a gigantic undertaking, I imagine, his and Bentham's ' Genera Plantarum ' will be ! I hope he will not get too much immersed in it, so as not to spare some time for Geographical Distribution and other such questions.

I have begun to work steadily, but very slowly as usual, at details on variation under domestication.

My dear Gray,

Yours always truly and gratefully,

C. DARWIN.

C. Darwin to C. Lyell.

Down [May 8th, 1860].

. . . . . . I have sent for the ' Canadian Naturalist.' If I cannot procure a copy I will borrow yours. I had a letter from Henslow this morning, who says that Sedgwick was, on last Monday night, to open a battery on me at the Cambridge Philosophical Society. Anyhow, I am much honoured by being attacked there, and at the Royal Society of Edinburgh.

I do not think it worth while to contradict single cases, nor is it worth while arguing against those who do not attend to what I state. A moment's reflection will show you that there must be (on our doctrine) large genera not varying (see p. 56 on the subject, in the second edition of the ' Origin'). Though I do not there discuss the case in detail.

It may be sheer bigotry for my own notions, but I prefer to the Atlantis, my notion of plants and animals having migrated from the Old to the New World, or conversely, when the climate was much hotter, by approximately the line of Behring's Straits. It is most important, as you say, to see living forms of plants going back so far in time. I wonder whether we shall ever discover the flora of the dry land of the coal period, and find it not so anomalous as the swamp or coal-making flora. I am working away over the blessed

Pigeon Manuscript; but, from one cause or another, I get on very slowly. . . .

This morning I got a letter from the Academy of Natural Sciences of Philadelphia, announcing that I am elected a correspondent. . . . It shows that some Naturalists there do not think me such a scientific profligate as many think me here.

My dear Lyell, yours gratefully,

C. DARWIN.

P.S.—What a grand fact about the extinct stag's horn worked by man!

C. Darwin to J. D. Hooker.

Down [May 13th, 1860].

MY DEAR HOOKER,—I return Henslow, which I was very glad to see. How good of him to defend me.* I will write and thank him.

As you said you were curious to hear Thomson's† opinion, I send his kind letter. He is evidently a strong opposer to us.

Yours affectionately,

C. DARWIN.

C. Darwin to J. D. Hooker.

Down [May 15th, 1860].

. . . . . . How paltry it is in such men as X., Y. and Co. not reading your essay. It is incredibly paltry.‡ They may all attack me to their hearts' content. I am got case-hardened. As for the old fogies in Cambridge, it really signifies nothing. I look at their attacks as a proof that our work is worth the doing. It makes me resolve to buckle on my

---

* Against Sedgwick's attack before the Cambridge Philosophical Society.

† Dr. Thomas Thomson, the Indian botanist. He was a collaborateur in Hooker and Thomson's 'Flora Indica,' 1855.

‡ These remarks do not apply to Dr. Harvey, who was, however, in a somewhat similar position. See p. 313.

armour. I see plainly that it will be a long uphill fight.
But think of Lyell's progress with Geology. One thing I
see most plainly, that without Lyell's, yours, Huxley's, and
Carpenter's aid, my book would have been a mere flash in
the pan. But if we all stick to it, we shall surely gain the
day. And I now see that the battle is worth fighting. I
deeply hope that you think so. Does Bentham progress
at all? I do not know what to say about Oxford. *
I should like it much with you, but it must depend on
health. . . .

Yours most affectionately,

C. DARWIN.

C. Darwin to C. Lyell.

Down, May 18th [1860].

MY DEAR LYELL,—I send a letter from Asa Gray to show
how hotly the battle rages there. Also one from Wallace,
very just in his remarks, though too laudatory and too modest,
and how admirably free from envy or jealousy. He must be
a good fellow. Perhaps I will enclose a letter from Thomson
of Calcutta; not that it is much, but Hooker thinks so highly
of him. . . .

Henslow informs me that Sedgwick† and then Professor
Clarke [*sic*]‡ made a regular and savage onslaught on my
book lately at the Cambridge Philosophical Society, but
Henslow seems to have defended me well, and maintained
that the subject was a legitimate one for investigation. Since

* His health prevented him from going to Oxford for the meeting of the British Association.

† Sedgwick's address is given somewhat abbreviated in *The Cambridge Chronicle*, May 19th, 1860.

‡ The late William Clark, Professor of Anatomy. My father seems to have misunderstood his informant. I am assured by Mr. J. W. Clark that his father (Prof. Clark) did not support Sedgwick in the attack.

then Phillips * has given lectures at Cambridge on the same subject, but treated it very fairly. How splendidly Asa Gray is fighting the battle. The effect on me of these multiplied attacks is simply to show me that the subject is worth fighting for, and assuredly I will do my best. . . . I hope all the attacks make you keep up your courage, and courage you assuredly will require. . . .

C. Darwin to A. R. Wallace.

Down, May 18th, 1860.

MY DEAR MR. WALLACE,—I received this morning your letter from Amboyna, dated February 16th, containing some remarks and your too high approval of my book. Your letter has pleased me very much, and I most completely agree with you on the parts which are strongest and which are weakest. The imperfection of the Geological Record is, as you say, the weakest of all ; but yet I am pleased to find that there are almost more geological converts than of pursuers of other branches of natural science. . . . I think geologists are more easily converted than simple naturalists, because more accustomed to reasoning. Before telling you about the progress of opinion on the subject, you must let me say how I admire the generous manner in which you speak of my book. Most persons would in your position have felt some envy or jealousy. How nobly free you seem to be of this common failing of mankind. But you speak far too modestly of yourself. You would, if you had my leisure, have done the work just as well, perhaps better, than I have done it. . . . .

* John Phillips, M.A., F.R.S., born 1800, died · 1874, from the effects of a fall. Professor of Geology at King's College, London, and afterwards at Oxford. He gave the 'Rede' lecture at Cambridge on May 15th, 1860, on 'The Succession of Life on the earth.' The Rede Lecturer is appointed annually by the Vice-Chancellor, and is paid by an endowment left in 1524 by Sir Robert Rede, Lord Chief Justice, in the reign of Henry VIII.

. . . Agassiz sends me a personal civil message, but incessantly attacks me ; but Asa Gray fights like a hero in defence. Lyell keeps as firm as a tower, and this autumn will publish on the 'Geological History of Man,' and will then declare his conversion, which now is universally known. I hope that you have received Hooker's splendid essay. . . . Yesterday I heard from Lyell that a German, Dr. Schaaffhausen,* has sent him a pamphlet published some years ago, in which the same view is nearly anticipated ; but I have not yet seen this pamphlet. My brother, who is a very sagacious man, always said, "you will find that some one will have been before you." I am at work at my larger work, which I shall publish in a separate volume. But from ill-health and swarms of letters, I get on very very slowly. I hope that I shall not have wearied you with these details. With sincere thanks for your letter, and with most deeply felt wishes for your success in science, and in every way, believe me,

Your sincere well-wisher,

C. DARWIN.

C. Darwin to Asa Gray.

Down, May 22nd [1860].

MY DEAR GRAY,—Again I have to thank you for one of your very pleasant letters of May 7th, enclosing a very pleasant remittance of £22. I am in simple truth astonished at all the kind trouble you have taken for me. I return Appletons' account. For the chance of your wishing for a formal acknowledgement I send one. If you have any further communication to the Appletons, pray express my acknowledgement for [their] generosity ; for it is generosity in my opinion. I am not at all surprised at the sale diminishing ; my extreme

---

* Hermann Schaaffhausen 'Ueber Beständigkeit und Umwandlung der Arten.' Verhandl. d. Naturhist. Vereins, Bonn, 1853. See 'Origin,' Historical Sketch.

surprise is at the greatness of the sale. No doubt the public has been *shamefully* imposed on! for they bought the book thinking that it would be nice easy reading. I expect the sale to stop soon in England, yet Lyell wrote to me the other day that calling at Murray's he heard that fifty copies had gone in the previous forty-eight hours. I am extremely glad that you will notice in 'Silliman' the additions in the 'Origin.' Judging from letters (and I have just seen one from Thwaites to Hooker), and from remarks, the most serious omission in my book was not explaining how it is, as I believe, that all forms do not necessarily advance, how there can now be *simple* organisms still existing. . . . I hear there is a *very* severe review on me in the 'North British,' by a Rev. Mr. Dunns,* a Free Kirk minister, and dabbler in Natural History. I should be very glad to see any good American reviews, as they are all more or less useful. You say that you shall touch on other reviews. Huxley told me some time ago that after a time he would write a review on all the reviews, whether he will I know not. If you allude to the 'Edinburgh,' pray notice *some* of the points which I will point out on a separate slip. In the *Saturday Review* (one of our cleverest periodicals) of May 5th, p. 573, there is a nice article on [the 'Edinburgh'] review, defending Huxley, but not Hooker; and the latter, I think, [the 'Edinburgh' reviewer] treats most ungenerously.† But surely you will get sick unto death of me and my reviewers.

With respect to the theological view of the question. This is always painful to me. I am bewildered. I had no inten-

---

* This statement as to authorship was made on the authority of Robert Chambers.

† In a letter to Mr. Huxley my father wrote : "Have you seen the last *Saturday Review*? I am very glad of the defence of you and of myself. I wish the reviewer had noticed Hooker. The reviewer, whoever he is, is a jolly good fellow, as this review and the last on me showed. He writes capitally, and understands well his subject. I wish he had slapped [the 'Edinburgh' reviewer] a little bit harder."

tion to write atheistically. But I own that I cannot see as plainly as others do, and as I should wish to do, evidence of design and beneficence on all sides of us.   There seems to me too much misery in the world.   I cannot persuade myself that a beneficent and omnipotent God would have designedly created the Ichneumonidæ with the express intention of their feeding within the living bodies of Caterpillars, or that a cat should play with mice.   Not believing this, I see no necessity in the belief that the eye was expressly designed.   On the other hand, I cannot anyhow be contented to view this wonderful universe, and especially the nature of man, and to conclude that everything is the result ·of brute force.   I am inclined to look at everything as resulting from designed laws, with the details, whether good or bad, left to the working out of what we may call chance.   Not that this notion *at all* satisfies me.   I feel most deeply that the whole subject is too profound for the human intellect.   A dog might as well speculate on the mind of Newton.   Let each man hope and believe what he can.   Certainly I agree with you that my views are not at all necessarily atheistical.   The lightning kills a man, whether a good one or bad one, owing to the excessively complex action of natural ·laws.   A child (who may turn out an idiot) is born by the action of even more complex laws, and I can see no reason why a man, or other animal, may not have been aboriginally produced by other laws, and that all these laws may have been expressly designed by an omniscient Creator, who foresaw every future event and consequence.   But the more I think the more bewildered I become ; as indeed I have probably shown by this letter.

Most deeply do I feel your generous kindness and interest.

Yours sincerely and cordially,

CHARLES DARWIN.

[Here follow my father's criticisms on the 'Edinburgh Review':

"What a quibble to pretend he did not understand what I meant by *inhabitants* of South America; and any one would suppose that I had not throughout my volume touched on Geographical Distribution. He ignores also everything which I have said on Classification, Geological Succession, Homologies, Embryology, and Rudimentary Organs—p. 496.

He falsely applies what I said (too rudely) about "blindness of preconceived opinions" to those who believe in creation, whereas I exclusively apply the remark to those who give up multitudes of species as true species, but believe in the remainder—p. 500.

He slightly alters what I say,—I *ask* whether creationists really believe that elemental atoms have flashed into life. He says that I describe them as so believing, and this, surely, is a difference—p. 501.

He speaks of my "clamouring against" all who believe in creation, and this seems to me an unjust accusation—p. 501.

He makes me say that the dorsal vertebræ vary; this is simply false: I nowhere say a word about dorsal vertebræ—p. 522.

What an illiberal sentence that is about my pretension to candour, and about my rushing through barriers which stopped Cuvier: such an argument would stop any progress in science—p. 525.

How disingenuous to quote from my remark to you about my *brief* letter [published in the 'Linn. Soc. Journal'], as if it applied to the whole subject—p. 530.

How disingenuous to say that we are called on to accept the theory, from the imperfection of the geological record, when I over and over again [say] how grave a difficulty the imperfection offers—p. 530."]

C. Darwin to J. D. Hooker.

Down, May 30th [1860].

MY DEAR HOOKER,—I return Harvey's letter, I have been very glad to see the reason why he has not read your Essay.

I feared it was bigotry, and I am glad to see that he goes a little way (*very much* further than I supposed) with us. . . .

I was not sorry for a natural opportunity of writing to Harvey, just to show that I was not piqued at his turning me and my book into ridicule,* not that I think it was a proceeding which I deserved, or worthy of him. It delights me that you are interested in watching the progress of opinion on the change of Species; I feared that you were weary of the subject; and therefore did not send A. Gray's letters. The battle rages furiously in the United States. Gray says he was preparing a speech, which would take $1\frac{1}{2}$ hours to deliver, and which he "fondly hoped would be a stunner." He is fighting splendidly, and there seem to have been many discussions with Agassiz and others at the meetings. Agassiz pities me much at being so deluded. As for the progress of opinion, I clearly see that it will be excessively slow, almost as slow as the change of species. . . . I am getting wearied at the storm of hostile reviews and hardly any useful. . . .

C. Darwin to C. Lyell.

Down, Friday night [June 1st, 1860].

. . . Have you seen Hopkins † in the new 'Fraser'? the public will, I should think, find it heavy. He will be dead

---

* A "serio-comic squib," read before the 'Dublin University Zoological and Botanical Association,' Feb. 17, 1860, and privately printed. My father's presentation copy is inscribed, "With the writer's *repentance*, Oct. 1860."

† William Hopkins died in 1866, "in his seventy-third year." He began life with a farm in Suffolk, but ultimately entered, comparatively late in life, at Peterhouse, Cambridge; he took his degree in 1827, and afterwards became an Esquire Bedell of the University. He was chiefly known as a mathematical "coach," and was eminently successful in the manufacture of Senior Wranglers. Nevertheless Mr. Stephen says ('Life of Fawcett,' p. 26) that he "was conspicuous for inculcating" a "liberal view of the studies of the place. He endeavoured to stimulate a philosophical interest in the mathematical sciences, instead of simply rousing

against me as you prophesied; but he is generously civil to me personally.* On his standard of proof, *natural* science would never progress, for without the making of theories I am convinced there would be no observation.

. . . . I have begun reading the 'North British,'† which so far strikes me as clever.

Phillips's Lecture at Cambridge is to be published.

All these reiterated attacks will tell heavily; there will be no more converts, and probably some will go back. I hope you do not grow disheartened, I am determined to fight to the last. I hear, however, that the great Buckle highly approves of my book.

I have had a note from poor Blyth, ‡ of Calcutta, who is

---

an ardour for competition." He contributed many papers on geological and mathematical subjects to the scientific journals. He had a strong influence for good over the younger men with whom he came in contact. The letter which he wrote to Henry Fawcett on the occasion of his blindness illustrates this. Mr. Stephen says ('Life of Fawcett,' p. 48) that by " this timely word of good cheer," Fawcett was roused from " his temporary prostration," and enabled to take a " more cheerful and resolute tone."

* 'Fraser's Magazine,' June 1860. My father, no doubt, refers to the following passage, p. 752, where the Reviewer expresses his " full participation in the high respect in which the author is universally held, both as a man and a naturalist; and the more so, because in the remarks which will follow in the second part of this Essay we shall be found to differ widely from him as regards many of his conclusions and the reasonings on which he

has founded them, and shall claim the full right to express such differences of opinion with all that freedom which the interests of scientific truth demands, and which we are sure Mr. Darwin would be one of the last to refuse to any one prepared to exercise it with candour and courtesy." Speaking of this review, my father wrote to Dr. Asa Gray : " I have remonstrated with him [Hopkins] for so coolly saying that I base my views on what I reckon as great difficulties. Any one, by taking these difficulties alone, can make a most strong case against me. I could myself write a more damning review than has as yet appeared !" A second notice by Hopkins appeared in the July number of ' Fraser's Magazine.'

† May 1860.

‡ Edward Blyth, born 1810, died 1873. His indomitable love of natural history made him neglect the druggist's business with which he started in life, and he soon got into serious difficulties. After sup-

much disappointed at hearing that Lord Canning will not grant any money; so I much fear that all your great pains will be thrown away. Blyth says (and he is in many respects a very good judge) that his ideas on Species are quite revolutionized . . . .

C. Darwin to J. D. Hooker.

Down, June 5th [1860].

MY DEAR HOOKER,—It is a pleasure to me to write to you, as I have no one to talk about such matters as we write on. But I seriously beg you not to write to me unless so inclined; for busy as you are, and seeing many people, the case is very different between us. . . .

Have you seen —— s abusive article on me? . . . It outdoes even the 'North British' and 'Edinburgh' in misapprehension and misrepresentation. I never knew anything so unfair as in discussing cells of bees, his ignoring the case of Melipona, which builds combs almost exactly intermediate between hive and humble bees. What has —— done that he feels so immeasurably superior to all us wretched naturalists, and to all political economists, including that great philosopher Malthus? This review, however, and Harvey's letter have convinced me that I must be a very bad explainer. Neither

---

porting himself for a few years as a writer on Field Natural History, he ultimately went out to India as Curator of the Museum of the R. Asiatic Soc. of Bengal, where the greater part of his working life was spent. His chief publications were the monthly reports made as part of his duty to the Society. He had stored in his remarkable memory a wonderful wealth of knowledge, especially with regard to the mammalia and birds of India—knowledge of which he freely gave to those who asked. His letters to my father give evidence of having been carefully studied, and the long list of entries after his name in the index to 'Animals and Plants,' show how much help was received from him. His life was an unprosperous and unhappy one, full of money difficulties and darkened by the death of his wife after a few years of marriage.

really understand what I mean by Natural Selection. I am inclined to give up the attempt as hopeless. Those who do not understand, it seems, cannot be made to understand.

By the way, I think, we entirely agree, except perhaps that I use too forcible language about selection. I entirely agree, indeed would almost go further than you when you say that climate (*i.e.* variability from all unknown causes) is "an active handmaid, influencing its mistress most materially." Indeed, I have never hinted that Natural Selection is "the efficient cause to the exclusion of the other," *i.e.* variability from Climate, &c. The very term *selection* implies something, *i.e.* variation or difference, to be selected. . . .

How does your book progress (I mean your general sort of book on plants), I hope to God you will be more successful than I have been in making people understand your meaning. I should begin to think myself wholly in the wrong, and that I was an utter fool, but then I cannot yet persuade myself, that Lyell, and you and Huxley, Carpenter, Asa Gray, and Watson, &c., are all fools together. Well, time will show, and nothing but time. Farewell. . . .

C. Darwin to C. Lyell.

Down, June 6th [1860].

. . . It consoles me that —— sneers at Malthus, for that clearly shows, mathematician though he may be, he cannot understand common reasoning. By the way what a discouraging example Malthus is, to show during what long years the plainest case may be misrepresented and misunderstood. I have read the ' Future'; how curious it is that several of my reviewers should advance such wild arguments, as that varieties of dogs and cats do not mingle ; and should bring up the old exploded doctrine of definite analogies . . . I am beginning to despair of ever making the majority understand my notions. Even Hopkins does not thoroughly. By the way, I have been so much pleased by the way he personally alludes to me. I must be a very bad explainer. I hope to Heaven that you will succeed better. Several reviews and several letters have shown me too clearly how little I am understood. I suppose "natural selection" was a bad term; but to change it now, I think, would make confusion worse confounded, nor can I think of a better; "Natural Preservation" would not imply a preservation of particular varieties, and would seem a truism, and would not bring man's and nature's selection under one point of view. I can only hope by reiterated explanations finally to make the matter clearer. If my MS. spreads out, I think I shall publish one volume exclusively on variation of animals and plants under domestication. I want to show that I have not been quite so rash as many suppose.

Though weary of reviews, I should like to see Lowell's * some time. . . . I suppose Lowell's difficulty about instinct is the same as Bowen's; but it seems to me wholly to rest on the assumption that instincts cannot graduate as finely as structures. I have stated in my volume that it is hardly possible to know which, *i.e.* whether instinct or structure, change first by insensible steps. Probably sometimes instinct, sometimes structure. When a British insect feeds on an exotic plant, instinct has changed by very small steps, and their structures might change so as to fully profit by the new food. Or structure might change first, as the direction of tusks in one variety of Indian elephants, which leads it to attack the tiger in a different manner from other kinds of elephants. Thanks for your letter of the 2nd, chiefly about Murray. (N.B. Harvey of Dublin gives me, in a letter, the argument of tall men marrying short women, as one of great weight ! †)

---

* The late J. A. Lowell in the 'Christian Examiner' (Boston, U. S.), May, 1860.

† See footnote, *ante*, p. 261.

I do not quite understand what you mean by saying, "that the more they prove that you underrate physical conditions, the better for you, as Geology comes in to your aid."

. . . I see in Murray and many others one incessant fallacy, when alluding to slight differences of physical conditions as being very important; namely, oblivion of the fact that all species, except very local ones, range over a considerable area, and though exposed to what the world calls considerable *diversities*, yet keep constant. I have just alluded to this in the 'Origin' in comparing the productions of the Old and the New Worlds. Farewell, shall you be at Oxford? If H. gets quite well, perhaps I shall go there.

Yours affectionately,
C. DARWIN.

C. Darwin to C. Lyell.

Down [June 14th, 1860].

. . . Lowell's review * is pleasantly written, but it is clear that he is not a naturalist. He quite overlooks the importance of the accumulation of mere individual differences, and which, I think I can show, is the great agency of change under domestication. I have not finished Schaaffhausen, as I read German so badly. I have ordered a copy for myself, and should like to keep yours till my own arrives, but will return it to you instantly if wanted. He admits statements rather rashly, as I dare say I do. I see only one sentence as yet at all approaching natural selection.

There is a notice of me in the penultimate number of 'All the Year Round,' but not worth consulting; chiefly a well-done hash of my own words. Your last note was very interesting and consolatory to me.

I have expressly stated that I believe physical conditions have a more direct effect on plants than on animals. But the

---

* J. A. Lowell in the 'Christian Examiner,' May 1860.

more I study, the more I am led to think that natural selection regulates, in a state of nature, most trifling differences. As squared stone, or bricks, or timber, are the indispensable materials for a building, and influence its character, so is variability not only indispensable but influential. Yet in the same manner as the architect is the *all* important person in a building, so is selection with organic bodies. . . . .

[The meeting of the British Association at Oxford in 1860 is famous for two pitched battles over the 'Origin of Species.' Both of them originated in unimportant papers. On Thursday, June 28, Dr. Daubeny of Oxford made a communication to Section D : " On the final causes of the sexuality of plants, with particular reference to Mr. Darwin's work on the 'Origin of Species.' " Mr. Huxley was called on by the President, but tried (according to the *Athenæum* report) to avoid a discussion, on the ground "that a general audience, in which sentiment would unduly interfere with intellect, was not the public before which such a discussion should be carried on." However, the subject was not allowed to drop. Sir R. Owen (I quote from the *Athenæum*, July 7, 1860), who "wished to approach this subject in the spirit of the philosopher," expressed his " conviction that there were facts by which the public could come to some conclusion with regard to the probabilities of the truth of Mr. Darwin's theory." He went on to say that the brain of the gorilla " presented more differences, as compared with the brain of man, than it did when compared with the brains of the very lowest and most problematical of the Quadrumana." Mr. Huxley replied, and gave these assertions a " direct and unqualified contradiction," pledging himself to " justify that unusual procedure elsewhere," * a pledge which he amply fulfilled.† On Friday there was peace, but on Saturday 30th, the battle arose with

* 'Man's Place in Nature,' by   † See the 'Nat. Hist. Review,'
T. H. Huxley, 1863, p. 114.   1861.

redoubled fury over a paper by Dr. Draper of New York, on the 'Intellectual development of Europe considered with reference to the views of Mr. Darwin.'

The following account is from an eye-witness of the scene.

"The excitement was tremendous. The Lecture-room, in which it had been arranged that the discussion should be held, proved far too small for the audience, and the meeting adjourned to the Library of the Museum, which was crammed to suffocation long before the champions entered the lists. The numbers were estimated at from 700 to 1000. Had it been term-time, or had the general public been admitted, it would have been impossible to have accommodated the rush to hear the oratory of the bold Bishop. Professor Henslow, the President of Section D, occupied the chair, and wisely announced *in limine* that none who had not valid arguments to bring forward on one side or the other, would be allowed to address the meeting : a caution that proved necessary, for no fewer than four combatants had their utterances burked by him, because of their indulgence in vague declamation.

"The Bishop was up to time, and spoke for full half-an-hour with inimitable spirit, emptiness and unfairness. It was evident from his handling of the subject that he had been 'crammed' up to the throat, and that he knew nothing at first hand ; in fact, he used no argument not to be found in his 'Quarterly' article. He ridiculed Darwin badly, and Huxley savagely, but all in such dulcet tones, so persuasive a manner, and in such well-turned periods, that I who had been inclined to blame the President for allowing a discussion that could serve no scientific purpose, now forgave him from the bottom of my heart. Unfortunately the Bishop, hurried along on the current of his own eloquence, so far forgot himself as to push his attempted advantage to the verge of personality in a telling passage in which he turned round and addressed Huxley : I forget the precise words, and quote from Lyell. 'The Bishop asked whether Huxley was related by his grand-

father's or grandmother's side to an ape.'* Huxley replied to
the scientific argument of his opponent with force and eloquence, and to the personal allusion with a self-restraint, that
gave dignity to his crushing rejoinder."

Many versions of Mr. Huxley's speech were current: the
following report of his conclusion is from a letter addressed
by the late John Richard Green, then an undergraduate, to
a fellow-student, now Professor Boyd Dawkins. " I asserted,
and I repeat, that a man has no reason to be ashamed of
having an ape for his grandfather. If there were an ancestor
whom I should feel shame in recalling, it would be a *man*, a
man of restless and versatile intellect, who, not content with
an equivocal† success in his own sphere of activity, plunges
into scientific questions with which he has no real acquaintance, only to obscure them by an aimless rhetoric, and dis-
tract the attention of his hearers from the real point at issue
by eloquent digressions, and skilled appeals to religious
prejudice." ‡

The letter above quoted continues :

" The excitement was now at its height ; a lady fainted and
had to be carried out, and it was some time before the discussion was resumed. Some voices called for Hooker, and
his name having been handed up, the President invited him to
give his view of the theory from the Botanical side. This he
did, demonstrating that the Bishop, by his own showing, had
never grasped the principles of the ' Origin,' and that he was
absolutely ignorant of the elements of botanical science. The
Bishop made no reply, and the meeting broke up.

" There was a crowded conversazione in the evening at the

---

* Lyell's ' Letters,' vol. ii. p. 335.

† Professor Victor Carus, who
has a distinct recollection of the
scene, does not remember the word
equivocal. He believes, too, that
Lyell's version of the ape sentence
is slightly incorrect.

‡ Mr. Fawcett wrote (' Macmillan's Magazine,' 1860) :
" The retort was so justly deserved
and so inimitable in its manner,
that no one who was present can
ever forget the impression that it
made."

rooms of the hospitable and genial Professor of Botany, Dr. Daubeny, where the almost sole topic was the battle of the 'Origin,' and I was much struck with the fair and unprejudiced way in which the black coats and white cravats of Oxford discussed the question, and the frankness with which they offered their congratulations to the winners in the combat."]

C. Darwin to J. D. Hooker.

Sudbrook Park, Monday night
[July 2nd, 1860].

MY DEAR HOOKER,—I have just received your letter. I have been very poorly, with almost continuous bad headache for forty-eight hours, and I was low enough, and thinking what a useless burthen I was to myself and all others, when your letter came, and it has so cheered me; your kindness and affection brought tears into my eyes. Talk of fame, honour, pleasure, wealth, all are dirt compared with affection; and this is a doctrine with which, I know, from your letter, that you will agree with from the bottom of your heart. . . . How I should have liked to have wandered about Oxford with you, if I had been well enough; and how still more I should have liked to have heard you triumphing over the Bishop. I am astonished at your success and audacity. It is something unintelligible to me how any one can argue in public like orators do. I had no idea you had this power. I have read lately so many hostile views, that I was beginning to think that perhaps I was wholly in the wrong, and that —— was right when he said the whole subject would be forgotten in ten years; but now that I hear that you and Huxley will fight publicly (which I am sure I never could do), I fully believe that our cause will, in the long-run, prevail. I am glad I was not in Oxford, for I should have been overwhelmed, with my [health] in its present state.

C. Darwin to T. H. Huxley.

Sudbrook Park, Richmond,
July 3rd (1860).

. . . . I had a letter from Oxford, written by Hooker late
on Sunday night, giving me some account of the awful battles
which have raged about species at Oxford. He tells me you
fought nobly with Owen (but I have heard no particulars),
and that you answered the B. of O. capitally. I often think
that my friends (and you far beyond others) have good cause
to hate me, for having stirred up so much mud, and led them
into so much odious trouble. If I had been a friend of
myself, I should have hated me. (How to make that sentence
good English, I know not.) But remember, if I had not
stirred up the mud, some one else certainly soon would. I
honour your pluck ; I would as soon have died as tried to
answer the Bishop in such an assembly. . . .

[On July 20th, my father wrote to Mr. Huxley:

"From all that I hear from several quarters, it seems that
Oxford did the subject great good. It is of enormous importance, the showing the world that a few first-rate men are
not afraid of expressing their opinion."]

C. Darwin to J. D. Hooker.

[July 1860.]

. . . . I have just read the 'Quarterly.'* It is uncommonly clever; it picks out with skill all the most conjectural

* 'Quarterly Review,' July 1860.
The article in question was by
Wilberforce, Bishop of Oxford, and
was afterwards published in his
" Essays Contributed to the 'Quarterly Review,' 1874." The passage
from the 'Anti-Jacobin' gives the
history of the evolution of space
from the " primæval point or
punctum saliens of the universe,"

parts, and brings forward well all the difficulties. It quizzes me quite splendidly by quoting the 'Anti-Jacobin' versus my Grandfather. You are not alluded to, nor, strange to say, Huxley; and I can plainly see, here and there, ——'s hand. The concluding pages will make Lyell shake in his shoes. By Jove, if he sticks to us, he will be a real hero. Good-night. Your well-quizzed, but not sorrowful, and affectionate friend.   C. D.

I can see there has been some queer tampering with the Review, for a page has been cut out and reprinted.

----

which is conceived to have moved "forward in a right line, *ad infinitum*, till it grew tired ; after which the right line, which it had generated, would begin to put itself in motion in a lateral direction, describing an area of infinite extent. This area, as soon as it became conscious of its own existence, would begin to ascend or descend according as its specific gravity would determine it, forming an immense solid space filled with vacuum, and capable of containing the present universe."

The following (p. 263) may serve as an example of the passages in which the reviewer refers to Sir Charles Lyell :—"That Mr. Darwin should have wandered from this broad highway of nature's works into the jungle of fanciful assumption is no small evil. We trust that he is mistaken in believing that he may count Sir C. Lyell as one of his converts. We know, indeed, the strength of the temptations which he can bring to bear upon his geological brother. . . . Yet no man has been more distinct and more logical in the denial of the transmutation of species than Sir C. Lyell, and that not in the infancy of his scientific life, but in its full vigour and maturity." The Bishop goes on to appeal to Lyell, in order that with his help "this flimsy speculation may be as completely put down as was what in spite of all denials we must venture to call its twin though less instructed brother, the 'Vestiges of Creation.'"

With reference to this article, Mr. Brodie Innes, my father's old friend and neighbour, writes :— "Most men would have been annoyed by an article written with the Bishop's accustomed vigour, a mixture of argument and ridicule. Mr. Darwin was writing on some parish matter, and put a postscript —'If you have not seen the last 'Quarterly,' do get it ; the Bishop of Oxford has made such capital fun of me and my grandfather. By a curious coincidence, when I received the letter, I was staying in the same house with the Bishop, and showed it to him. He said, 'I am very glad he takes it in that way, he is such a capital fellow.'"

[Writing on July 22 to Dr. Asa Gray my father thus refers to Lyell's position :—

"Considering his age, his former views and position in society, I think his conduct has been heroic on this subject."]

C. Darwin to Asa Gray.

[Hartfield, Sussex] July 22nd [1860].

MY DEAR GRAY,—Owing to absence from home at water-cure and then having to move my sick girl to whence I am now writing, I have only lately read the discussion in Proc. American Acad.,* and now I cannot resist expressing my sincere admiration of your most clear powers of reasoning. As Hooker lately said in a note to me, you are more than *any one* else the thorough master of the subject. I declare that you know my book as well as I do myself; and bring to the question new lines of illustration and argument in a manner which excites my astonishment and almost my envy! I admire these discussions, I think, almost more than your article in Silliman's Journal. Every single word seems weighed carefully, and tells like a 32-pound shot. It makes me much wish (but I know that you have not time) that you could write more in detail, and give, for instance, the facts on the variability of the American wild fruits. The *Athenæum* has the largest circulation, and I have sent my copy to the editor with a request that he would republish the first discussion ; I much fear he will not, as he reviewed the subject in so hostile a spirit . . . I shall be curious (and will order) the August number, as soon as I know that it contains your review of Reviews. My conclusion is that you have made a mistake in being a botanist, you ought to have been a lawyer.

* April 10, 1860. Dr. Gray criticised in detail " several of the positions taken at the preceding meeting by Mr. [J. A.] Lowell, Prof. Bowen and Prof. Agassiz." It was reprinted in the *Athenæum*, Aug. 4, 1860.

. . . . Henslow * and Daubeny are shaken. I hear from Hooker that he hears from Hochstetter that my views are making very considerable progress in Germany, and the good workers are discussing the question. Bronn at the end of his translation has a chapter of criticism, but it is such difficult German that I have not yet read it. Hopkins's review in 'Fraser' is thought the best which has appeared against us. I believe that Hopkins is so much opposed because his course of study has never led him to reflect much on such subjects as geographical distribution, classification, homologies, &c., so that he does not feel it a relief to have some kind of explanation.

C. Darwin to C. Lyell.

Hartfield [Sussex], July 30th [1860].

. . . . I had lots of pleasant letters about the Brit. Assoc., and our side seems to have got on very well. There has been as much discussion on the other side of the Atlantic as on this. No one I think understands the whole case better than Asa Gray, and he has been fighting nobly. He is a capital reasoner. I have sent one of his printed discussions to our *Athenæum*, and the editor says he will print it. The 'Quarterly' has been out some time. It contains no malice, which is wonderful. . . . It makes me say many things which

---

* Professor Henslow was mentioned in the December number of 'Macmillan's Magazine' as being an adherent of Evolution. In consequence of this he published, in the February number of the following year, a letter defining his position. This he did by means of an extract from a letter addressed to him by the Rev. L. Jenyns (Blomefield) which "very nearly," as he says, expressed his views. Mr. Blomefield wrote, "I was not aware that you had become a convert to his (Darwin's) theory, and can hardly suppose you have accepted it as a whole, though, like myself, you may go to the length of imagining that many of the smaller groups, both of animals and plants, may at some remote period have had a common parentage. I do not with some say that the whole of his theory cannot be true—but that it is very far from proved; and I doubt its ever being possible to prove it."

I do not say. At the end it quotes all your conclusions against Lamarck, and makes a solemn appeal to you to keep firm in the true faith. I fancy it will make you quake a little. ——— has ingeniously primed the Bishop (with Murchison) against you as head of the uniformitarians. The only other review worth mentioning, which I can think of, is in the third No. of the 'London Review,' by some geologist, and favourable for a wonder. It is very ably done, and I should like much to know who is the author. I shall be very curious to hear on your return whether Bronn's German translation of the ' Origin ' has drawn any attention to the subject. Huxley is eager about a ' Natural History Review,' which he and others are going to edit, and he has got so many first-rate assistants, that I really believe he will make it a first-rate production. I have been doing nothing, except a little botanical work as amusement. I shall hereafter be very anxious to hear how your tour has answered. I expect your book on the geological history of Man will, with a vengeance, be a bomb-shell. I hope it will not be very long delayed. Our kindest remembrances to Lady Lyell. This is not worth sending, but I have nothing better to say.

Yours affectionately,
C. DARWIN.

C. Darwin to F. Watkins. *

Down, July 30th, [1860].

MY DEAR WATKINS,—Your note gave me real pleasure. Leading the retired life which I do, with bad health, I oftener think of old times than most men probably do ; and your face now rises before me, with the pleasant old expression, as vividly as if I saw you.

My book has been well abused, praised, and splendidly quizzed by the Bishop of Oxford ; but from what I see of its

* See Vol. I. p. 168.

influence on really good workers in science, I feel confident that, *in the main*, I am on the right road.   With respect to your question, I think the arguments are valid, showing that all animals have descended from four or five primordial forms ; and that analogy and weak reasons go to show that all have descended from some single prototype.

Farewell, my old friend.   I look back to old Cambridge days with unalloyed pleasure.

Believe me, yours most sincerely,
CHARLES DARWIN.

T. H. Huxley to C. Darwin.

August 6th, 1860.

My DEAR DARWIN,—I have to announce a new and great ally for you. . . . .

Von Bär writes to me thus :—" Et outre cela, je trouve que vous écrivez encore des rédactions.   Vous avez écrit sur l'ouvrage de M. Darwin une critique dont je n'ai trouvé que des débris dans un journal allemand.   J'ai oublié le nom terrible du journal anglais dans lequel se trouve votre récension.   En tout cas aussi je ne peux pas trouver le journal ici.   Comme je m'intéresse beaucoup pour les idées de M. Darwin, sur lesquelles j'ai parlé publiquement et sur lesquelles je ferai peut-être imprimer quelque chose—vous m'obligeriez infiniment si vous pourriez me faire parvenir ce que vous avez écrit sur ces idées.

" J'ai énoncé les mêmes idées sur la transformation des types ou origine d'espèces que M. Darwin.*   Mais c'est seulement sur la géographie zoologique que je m'appuie.   Vous trouverez, dans le dernier chapitre du traité ' Ueber Papuas und Alfuren,' que j'en parle très décidément sans savoir que M. Darwin s'occupait de cet objet."

The treatise to which Von Bär refers he gave me when over

---

* See footnote, Vol. II. p. 186.

here, but I have not been able to lay hands on it since this letter reached me two days ago. When I find it I will let you know what there is in it.

Ever yours faithfully,

T. H. HUXLEY.

C. Darwin to T. H. Huxley.

Down, August 8 [1860].

MY DEAR HUXLEY—Your note contained magnificent news, and thank you heartily for sending it me. Von Baer weighs down with a vengeance all the virulence of [the 'Edinburgh' reviewer] and weak arguments of Agassiz. If you write to Von Baer, for heaven's sake tell him that we should think one nod of approbation on our side, of the greatest value; and if he does write anything, beg him to send us a copy, for I would try and get it translated and published in the *Athenæum* and in 'Silliman' to touch up Agassiz. . . . . Have you seen Agassiz's weak metaphysical and theological attack on the 'Origin' in the last 'Silliman'?* I would send it you, but apprehend it would be less trouble for you to look at it in London than return it to me. R. Wagner has sent me a German pamphlet,† giving an abstract of Agassiz's 'Essay on Classification,' " mit Rücksicht auf Darwins Ansichten," &c. &c. He won't go very "dangerous lengths," but thinks the truth lies half-way between Agassiz and the 'Origin.' As he goes thus far he will, nolens

---

* The 'American Journal of Science and Arts' (commonly called 'Silliman's Journal'), July 1860. Printed from advanced sheets of vol. iii. of 'Contributions to the Nat. Hist. of the U. S.' My father's copy has a pencilled "Truly" opposite the following passage:— "Unless Darwin and his followers succeed in showing that the struggle for life tends to something beyond favouring the existence of certain individuals over that of other individuals, they will soon find that they are following a shadow."

† 'Louis Agassiz's Prinzipien der Classification, &c., mit Rücksicht auf Darwins Ansichten. Separat-Abdruck aus den Göttingischen gelehrten Anzeigen,' 1860.

volens, have to go further. He says he is going to review me in [his] yearly Report. My good and kind agent for the propagation of the Gospel—*i. e.* the devil's gospel.

Ever yours,
C. DARWIN.

C. Darwin to C. Lyell.

Down, August 11th [1860].

. . . I have laughed at Woodward thinking that you were a man who could be influenced in your judgment by the voice of the public ; and yet after mortally sneering at him, I was obliged to confess to myself, that I had had fears, what the effect might be of so many heavy guns fired by great men. As I have (sent by Murray) a spare 'Quarterly Review,' I send it by this post, as it may amuse you. The Anti-Jacobin part amused me. It is full of errors, and Hooker is thinking of answering it. There has been a cancelled page ; I should like to know what gigantic blunder it contained. Hooker says that —— has played on the Bishop, and made him strike whatever note he liked ; he has wished to make the article as disagreeable to you as possible. I will send the *Athenæum* in a day or two.

As you wish to hear what reviews have appeared, I may mention that Agassiz has fired off a shot in the last 'Silliman,' not good at all, denies variations and rests on the perfection of Geological evidence. Asa Gray tells me that a very clever friend has been almost converted to our side by this review of Agassiz's . . . Professor Parsons * has published in the same 'Silliman' a speculative paper correcting my notions, worth nothing. In the 'Highland Agricultural Journal' there is a review by some Entomologist, not worth much. This is all that I can remember. . . . As Huxley says, the platoon firing must soon cease. Hooker and

* Theophilus Parsons, Professor of Law in Harvard University.

Huxley, and Asa Gray, I see, are determined to stick to the battle and not give in ; I am fully convinced that whenever you publish, it will produce a great effect on all *trimmers*, and on many others.   By the way I forgot to mention Daubeny's pamphlet,* very liberal and candid, but scientifically weak. I believe Hooker is going nowhere this summer ; he is excessively busy . . . He has written me many, most nice letters.   I shall be very curious to hear on your return some account of your Geological doings.   Talking of Geology, you used to be interested about the " pipes " in the chalk.   About three years ago a perfectly circular hole suddenly appeared in a flat grass field to everyone's astonishment, and was filled up with many waggon loads of earth ; and now two or three days ago, again it has circularly subsided about two feet more.   How clearly this shows what is still slowly going on. This morning I recommenced work, and am at dogs ; when I have written my short discussion on them, I will have it copied, and if you like, you can then see how the argument stands, about their multiple origin.   As you seemed to think this important, it might be worth your reading ; though I do not feel sure that you will come to the same probable conclusion that I have done.   By the way, the Bishop makes a very telling case against me, by accumulating several instances where I speak very doubtfully ; but this is very unfair, as in such cases as this of the dog, the evidence is and must be very doubtful. . . .

C. Darwin to Asa Gray.

Down, August 11 [1860].

MY DEAR GRAY,—On my return home from Sussex about a week ago, I found several articles sent by you.   The first

---

* ' Remarks on the final causes of the sexuality of plants with particular reference to Mr. Darwin's   work on the " Origin of Species." ' —Brit. Assoc. Report, 1860.

article, from the 'Atlantic Monthly,' I am very glad to
possess. By the way, the editor of the *Athenæum* * has
inserted your answer to Agassiz, Bowen, and Co., and when
I therein read them, I admired them even more than at first.
They really seemed to me admirable in their condensation,
force, clearness and novelty.

I am surprised that Agassiz did not succeed in writing
something better. How absurd that logical quibble—" if
species do not exist, how can they vary?" As if any one
doubted their temporary existence. How coolly he assumes
that there is some clearly defined distinction between individual differences and varieties. It is no wonder that a man
who calls identical forms, when found in two countries, distinct species, cannot find variation in nature. Again, how
unreasonable to suppose that domestic varieties selected by
man for his own fancy (p. 147) should resemble natural
varieties or species. The whole article seems to me poor ; it
seems to me hardly worth a detailed answer (even if I could
do it, and I much doubt whether I possess your skill in
picking out salient points and driving a nail into them), and
indeed you have already answered several points. Agassiz's
name, no doubt, is a heavy weight against us. . . .

If you see Professor Parsons, will you thank him for the
extremely liberal and fair spirit in which his Essay † is written.
Please tell him that I reflected much on the chance of favourable monstrosities (*i.e.* great and sudden variation) arising. I
have, of course, no objection to this, indeed it would be a great
aid, but I did not allude to the subject, for, after much labour,
I could find nothing which satisfied me of the probability of
such occurrences. There seems to me in almost every case
too much, too complex, and too beautiful adaptation, in every
structure to believe in its sudden production. I have alluded
under the head of beautifully hooked seeds to such possibility. Monsters are apt to be sterile, or *not* to transmit

* Aug. 4, 1860.   † 'Silliman's Journal,' July 1860.

monstrous peculiarities.   Look at the fineness of gradation in the shells of successive *sub-stages* of the same great formation ; I could give many other considerations which made me doubt such view.   It holds, to a certain extent, with domestic productions no doubt, where man preserves some abrupt change in structure.   It amused me to see Sir R. Murchison quoted as a judge of affinities of animals, and it gave me a cold shudder to hear of any one speculating about a true crustacean giving birth to a true fish ! *

Yours most truly,
C. DARWIN.

C. Darwin to C. Lyell.

Down, September 1st [1860].

MY DEAR LYELL,—I have been much interested by your letter of the 28th, received this morning.   It has *delighted* me, because it demonstrates that you have thought a good deal lately on Natural Selection.   Few things have surprised me more than the entire paucity of objections and difficulties new to me in the published reviews.   Your remarks are of a different stamp and new to me.   I will run through them, and make a few pleadings such as occur to me.

I put in the possibility of the Galapagos having been *continuously* joined to America, out of mere subservience to the many who believe in Forbes's doctrine, and did not see the danger of admission, about small mammals surviving there in such case.   The case of the Galapagos, from certain facts on littoral sea-shells (viz. Pacific Ocean and South American littoral species), in fact convinced me more than in any other case of other islands, that the Galapagos had never been

---

* Parsons, *loc. cit.* p. 5, speaking of Pterichthys and Cephalaspis, says:—" Now is it too much to infer from these facts that either of these animals, if a crustacean, was so nearly a fish that some of its ova may have become fish ; or, if itself a fish, was so nearly a crustacean that it may have been born from the ovum of a crustacean ? "

continuously united with the mainland; it was mere base subservience, and terror of Hooker and Co.

With respect to atolls, I think mammals would hardly survive *very long*, even if the main islands (for as I have said in the Coral Book, the outline of groups of atolls do not look like a former *continent*) had been tenanted by mammals, from the extremely small area, the very peculiar conditions, and the probability that during subsidence all or nearly all atolls have been breached and flooded by the sea many times during their existence as atolls.

I cannot conceive any existing reptile being converted into a mammal. From homologies I should look at it as certain that all mammals had descended from some single progenitor. What its nature was, it is impossible to speculate. More like, probably, the Ornithorhynchus or Echidna than any known form; as these animals combine reptilian characters (and in a lesser degree bird character) with mammalian. We must imagine some form as intermediate, as is Lepidosiren now, between reptiles and fish, between mammals and birds on the one hand (for they retain longer the same embryological character) and reptiles on the other hand. With respect to a mammal not being developed on any island, besides want of time for so prodigious a development, there must have arrived on the island the necessary and peculiar progenitor, having a character like the embryo of a mammal; and not an *already developed* reptile, bird or fish.

We might give to a bird the habits of a mammal, but inheritance would retain almost for eternity some of the bird-like structure, and prevent a new creature ranking as a true mammal.

I have often speculated on antiquity of islands, but not with your precision, or at all under the point of view of Natural Selection *not* having done what might have been anticipated. The argument of littoral Miocene shells at the Canary Islands is new to me. I was deeply impressed (from

the amount of the denudation) [with the] antiquity of St.
Helena, and its age agrees with the peculiarity of the flora.
With respect to bats at New Zealand (N.B. There are two or
three European bats in Madeira, and I think in the Canary
Islands) not having given rise to a group of non-volant bats,
it is, now you put the case, surprising ; more especially as
the genus of bats in New Zealand is very peculiar, and therefore has probably been long introduced, and they now speak
of Cretacean fossils there. But the first necessary step has to
be shown, namely, of a bat taking to feed on the ground, or
anyhow, and anywhere, except in the air. I am bound to
confess I do know one single such fact, viz. of an Indian species
killing frogs. Observe, that in my wretched Polar Bear case,
I do show the first step by which conversion into a whale
"would be easy," "would offer no difficulty"!! So with seals,
I know of no fact showing any the least incipient variation of
seals feeding on the shore. Moreover, seals wander much ;
I searched in vain, and could not find *one* case of any species
of seal confined to any islands. And hence wanderers would
be apt to cross with individuals undergoing any change on an
island, as in the case of land birds of Madeira and Bermuda.
The same remark applies even to bats, as they frequently
come to Bermuda from the mainland, though about 600 miles
distant. With respect to the Amblyrhynchus of the Galapagos, one may infer as probable, from marine habits being
so rare with Saurians, and from the terrestrial species being
confined to a few central islets, that its progenitor first arrived
at the Galapagos ; from what country it is impossible to say,
as its affinity I believe is not very clear to any known species.
The offspring of the terrestrial species was probably rendered
marine. Now in this case I do not pretend I can show
variation in habits ; but we have in the terrestrial species a
vegetable feeder (in itself a rather unusual circumstance),
largely on *lichens*, and it would not be a great change for
its offspring to feed first on littoral algæ and then on sub-

marine algæ. I have said what I can in defence, but yours is a good line of attack. We should, however, always remember that no change will ever be effected till a variation in the habits or structure or of both *chance* to occur in the right direction, so as to give the organism in question an advantage over other already established occupants of land or water, and this may be in any particular case indefinitely long. I am very glad you will read my dogs MS., for it will be important to me to see what you think of the balance of evidence. After long pondering on a subject it is often hard to judge. With hearty thanks for your most interesting letter. Farewell.

My dear old master,

C. DARWIN.

C. Darwin to J. D. Hooker.

Down, September 2nd [1860].

MY DEAR HOOKER,—I am astounded at your news received this morning. I am become such an old fogy that I am amazed at your spirit. For God's sake do not go and get your throat cut. Bless my soul, I think you must be a little insane. I must confess it will be a most interesting tour; and, if you get to the top of Lebanon, I suppose extremely interesting—you ought to collect any beetles under stones there; but the Entomologists are such slow coaches. I dare say no result could be made out of them. [They] have never worked the Alpines of Britain.

If you come across any Brine lakes, do attend to their minute flora and fauna; I have often been surprised how little this has been attended to.

I have had a long letter from Lyell, who starts ingenious difficulties opposed to Natural Selection, because it has not done more than it has. This is very good, as it shows that he has thoroughly mastered the subject; and shows he is in

nothing

earnest. Very striking letter altogether and it rejoices the cockles of my heart.

. . . . How I shall miss you, my best and kindest of friends. God bless you.

Yours ever affectionately,

C. DARWIN.

C. Darwin to Asa Gray.

Down, Sept. 10 [1860].

. . . . You will be weary of my praise, but it * does strike me as quite admirably argued, and so well and pleasantly written. Your many metaphors are inimitably good. I said in a former letter that you were a lawyer, but I made a gross mistake, I am sure that you are a poet. No, by Jove, I will tell you what you are, a hybrid, a complex cross of lawyer, poet, naturalist and theologian! Was there ever such a monster seen before?

I have just looked through the passages which I have marked as appearing to me extra good, but I see that they are too numerous to specify, and this is no exaggeration. My eye just alights on the happy comparison of the colours of the prism and our artificial groups. I see one little error of fossil *cattle* in South America.

It is curious how each one, I suppose, weighs arguments in a different balance: embryology is to me by far the strongest single class of facts in favour of change of forms, and not one, I think, of my reviewers has alluded to this. Variation not coming on at a very early age, and being inherited at not a very early corresponding period, explains, as it seems to me, the grandest of all facts in natural history, or rather in zoology, viz. the resemblance of embryos.

[Dr. Gray wrote three articles in the 'Atlantic Monthly' for

* Dr. Gray in the 'Atlantic Monthly' for July, 1860.

July, August, and October, which were reprinted as a pamphlet in 1861, and now form chapter iii. in 'Darwiniana' (1876), with the heading 'Natural Selection not inconsistent with Natural Theology.']

C. Darwin to C. Lyell.

Down, September 12th [1860].

MY DEAR LYELL,—I never thought of showing your letter to any one. I mentioned in a letter to Hooker that I had been much interested by a letter of yours with original objections, founded chiefly on Natural Selection not having done so much as might have been expected. . . . . In your letter just received, you have improved your case versus Natural Selection; and it would tell with the public (do not be tempted by its novelty to make it too strong); yet it seems to me, not *really* very killing, though I cannot answer your case, especially, why Rodents have not become highly developed in Australia. You must assume that they have inhabited Australia for a very long period, and this may or may not be the case. But I feel that our ignorance is so profound, why one form is preserved with nearly the same structure, or advances in organisation or even retrogrades, or becomes extinct, that I cannot put very great weight on the difficulty. Then, as you say often in your letter, we know not how many geological ages it may have taken to make any great advance in organisation. Remember monkeys in the Eocene formations: but I admit that you have made out an excellent objection and difficulty, and I can give only unsatisfactory and quite vague answers, such as you have yourself put; however, you hardly put weight enough on the absolute necessity of variations first arising in the right direction, *videlicet*, of seals beginning to feed on the shore.

I entirely agree with what you say about only one species of many becoming modified. I remember this struck me

Z 2

much when tabulating the varieties of plants, and I have a
discussion somewhere on this point. It is absolutely implied
in my ideas of classification and divergence that only one or
two species, of even large genera, give birth to new species;
and many whole genera become *wholly* extinct . . . . Please
see p. 341 of the 'Origin.' But I cannot remember that I
have stated in the 'Origin' the fact of only very few species
in each genus varying. You have put the view much better
in your letter. Instead of saying as I often have, that very
few species vary at the same time, I ought to have said, that
very few species of a genus *ever* vary so as to become modified;
for this is the fundamental explanation of classification, and
is shown in my engraved diagram. . . .

I quite agree with you on the strange and inexplicable fact
of Ornithorhynchus having been preserved, and Australian
Trigonia, or the Silurian Lingula. I always repeat to myself
that we hardly know why any one single species is rare or
common in the best-known countries. I have got a set of
notes somewhere on the inhabitants of fresh water; and it
is singular how many of these are ancient, or intermediate
forms; which I think is explained by the competition having
been less severe, and the rate of change of organic forms
having been slower in small confined areas, such as all the
fresh waters make compared with sea or land.

I see that you do allude in the last page, as a difficulty, to
Marsupials not having become Placentals in Australia; but
this I think you have no right at all to expect; for we ought
to look at Marsupials and Placentals as having descended
from some intermediate and lower form. The argument of
Rodents not having become highly developed in Australia
(supposing that they have long existed there) is much stronger.
I grieve to see you hint at the creation " of distinct successive
types, as well as of a certain number of distinct aboriginal
types." Remember, if you admit this, you give up the embryological argument (*the weightiest of all to me*), and the

morphological or homological argument. You cut my throat, and your own throat ; and I believe will live to be sorry for it. So much for species.

The striking extract which E. copied was your own writing ! ! in a note to me, many long years ago—which she copied and sent to Mme. Sismondi ; and lately my aunt, in sorting her letters, found E.'s and returned them to her. . . . . I have been of late shamefully idle, *i.e.* observing * instead of writing, and how much better fun observing is than writing.

Yours affectionately,

C. DARWIN.

C. Darwin to C. Lyell.

15 Marine Parade, Eastbourne,
Sunday [September 23rd, 1860].

MY DEAR LYELL,—I got your letter of the 18th just before starting here. You speak of saving me trouble in answering Never think of this, for I look at every letter of yours as an honour and pleasure, which is a pretty deal more than I can say of some of the letters which I receive. I have now one of 13 *closely written folio pages* to answer on species ! . . . .

I have a very decided opinion that all mammals must have descended from a *single* parent. Reflect on the multitude of details, very many of them of extremely little importance to their habits (as the number of bones of the head, &c., covering of hair, identical embryological development, &c. &c.). Now this large amount of similarity I must look at as certainly due to inheritance from a common stock. I am aware that some cases occur in which a similar or nearly similar organ has been acquired by independent acts of natural selection. But in most of such cases of these apparently so closely similar organs, some important homological difference may be detected. Please read p. 193, beginning, " The electric organs,"

---

* Drosera.

and trust me that the sentence, " In all these cases of two very
distinct species," &c. &c., was not put in rashly, for I went
carefully into every case.   Apply this argument to the whole
frame, internal and external, of mammifers, and you will see
why I think so strongly that all have descended from one
progenitor.   I have just re-read your letter, and I am not
perfectly sure that I understand your point.

I enclose two diagrams showing the sort of manner I *conjecture* that mammals have been developed.   I thought a little
on this when writing page 429, beginning, " Mr. Waterhouse."
(Please read the paragraph.)   I have not knowledge enough
to choose between these two diagrams.   If the brain of Marsupials in embryo closely resembles that of Placentals, I
should strongly prefer No. 2, and this agrees with the antiquity of Microlestes.   As a general rule I should prefer No. 1
diagram ; whether or not Marsupials have gone on being
developed, or rising in rank, from a very early period would
depend on circumstances too complex for even a conjecture.
Lingula has not risen since the Silurian epoch, whereas other
molluscs may have risen.

A, in the following diagrams, represents an unknown form,
probably intermediate between Mammals, Reptiles and Birds,
as intermediate as Lepidosiren now is between Fish and
Batrachians.   This unknown form is probably more closely
related to Ornithorhynchus than to any other known form.

I do not think that the multiple origin of dogs goes against
the single origin of man. . . . . All the races of man are so
infinitely closer together than to any ape, that (as in the case
of descent of all mammals from one progenitor), I should look
at all races of men as having certainly descended from one
parent.   I should look at it as probable that the races of men
were less numerous and less divergent formerly than now,
unless, indeed, some lower and more aberrant race even than
the Hottentot has become extinct.   Supposing, as I do for
one believe, that our dogs have descended from two or three

wolves, jackals, &c.; yet these have, on *our view*, descended from a single remote unknown progenitor. With domestic dogs the question is simply whether the whole amount of difference has been produced since man domesticated a single species; or whether part of the difference arises in the state

DIAGRAM I.

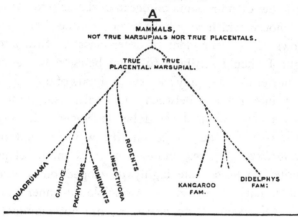

DIAGRAM II.

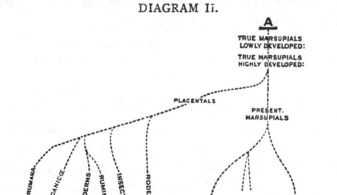

of nature. Agassiz and Co. think the negro and Caucasian are now distinct species, and it is a mere vain discussion whether, when they were rather less distinct, they would, on this standard of specific value, deserve to be called species.

I agree with your answer which you give to yourself on this point ; and the simile of man now keeping down any new man which might be developed, strikes me as good and new. The white man is " improving off the face of the earth " even races nearly his equals. With respect to islands, I think I would trust to want of time alone, and not to bats and rodents.

N.B.—I know of no rodents on oceanic islands (except my Galapagos mouse, which *may* have been introduced by man) keeping down the development of other classes. Still *much* more weight I should attribute to there being now, neither in islands nor elsewhere, [any] known animals of a grade of organisation intermediate between mammals, fish, reptiles, &c., whence a new mammal could be developed. If every vertebrate were destroyed throughout the world, except our *now well-established* reptiles, millions of ages might elapse before reptiles could become highly developed on a scale equal to mammals ; and, on the principle of inheritance, they would make some quite *new class*, and not mammals ; though *possibly* more intellectual ! I have not an idea that you will care for this letter, so speculative.

Most truly yours,
C. DARWIN.

C. Darwin to Asa Gray.

Down, Sept. 26 [1860].

. . . . I have had a letter of fourteen folio pages from Harvey against my book, with some ingenious and new remarks ; but it is an extraordinary fact that he does not understand at all what I mean by Natural Selection. I have begged him to read the Dialogue in next 'Silliman,' as you never touch the subject without making it clearer. I look at it as even more extraordinary that you never say a word or use an epithet which does not express fully my meaning. Now Lyell, Hooker, and others, who perfectly understand my book, yet sometimes use expressions to which I demur. Well, your extraordinary labour is over ; if there is any fair amount of truth in my view, I am well assured that your great labour has not been thrown away. . . .

I yet hope and almost believe, that the time will come when you will go further, in believing a very large amount of modification of species, than you did at first or do now. Can you tell me whether you believe further or more firmly than you did at first ? I should really like to know this. I can perceive in my immense correspondence with Lyell, who objected to much at first, that he has, perhaps unconsciously to himself, converted himself very much during the last six months, and I think this is the case even with Hooker. This fact gives me far more confidence than any other fact.

C. Darwin to C. Lyell.

15 Marine Parade, Eastbourne,
Friday evening [September 28th, 1860].

. . . . I am very glad to hear about the Germans reading my book. No one will be converted who has not independently begun to doubt about species. Is not Krohn * a good fellow ? I have long meant to write to him. He has been working at Cirripedes, and has detected two or three gigantic blunders, ... about which, I thank Heaven, I spoke rather doubtfully. Such difficult dissection that even Huxley failed. It is chiefly the interpretation which I put on parts that is so wrong, and not the parts which I describe. But they were gigantic blunders, and why I say all this is because Krohn, instead of crowing at all, pointed out my errors with the utmost gentleness and pleasantness. I have always

---

* There are two papers by Aug. Krohn, one on the Cement Glands, and the other on the development of Cirripedes, 'Wiegmann's Archiv,' xxv. and xxvi. My father has remarked that he " blundered dreadfully about the cement glands," ' Autobiography,' p. 81.

meant to write to him and thank him.   I suppose Dr. Krohn, Bonn, would reach him.

I cannot see yet how the multiple origin of dog can be properly brought as argument for the multiple origin of man. Is not your feeling a remnant of the deeply impressed one on all our minds, that a species is an entity, something quite distinct from a variety ?   Is it not that the dog case injures the argument from fertility, so that one main argument that the races of man are varieties and not species—*i.e.*, because they are fertile *inter se*, is much weakened ?

I quite agree with what Hooker says, that whatever variation is possible under culture, is *possible* under nature ; not that the same form would ever be accumulated and arrived at by selection for man's pleasure, and by natural selection for the organism's own good.

Talking of "natural selection ;" if I had to commence *de novo*, I would have used "natural preservation."   For I find men like Harvey of Dublin cannot understand me, though he has read the book twice.   Dr. Gray of the British Museum remarked to me that, "*selection* was obviously impossible with plants !   No one could tell him how it could be possible !" And he may now add that the author did not attempt it to him !

Yours ever affectionately,

C. DARWIN.

C. Darwin to C. Lyell.

15 Marine Parade, Eastbourne,
October 8th [1860].

MY DEAR LYELL,—I send the [English] translation of Bronn,* the first part of the chapter with generalities and praise is not translated.   There are some good hits.   He makes an apparently, and in part truly, telling case against me, says

---

* A MS. translation of Bronn's chapter of objections at the end of his German translation of the 'Origin of Species.'

that I cannot explain why one rat has a longer tail and another longer ears, &c. But he seems to muddle in assuming that these parts did not all vary together, or one part so insensibly before the other, as to be in fact contemporaneous. I might ask the creationist whether he thinks these differences in the two rats of any use, or as standing in some relation from laws of growth ; and if he admits this, selection might come into play. He who thinks that God created animals unlike for mere sport or variety, as man fashions his clothes, will not admit any force in my *argumentum ad hominem.*

Bronn blunders about my supposing several Glacial periods, whether or no such ever did occur.

He blunders about my supposing that development goes on at the same rate in all parts of the world. I presume that he has misunderstood this from the supposed migration into all regions of the more dominant forms.

I have ordered Dr. Bree,* and will lend it to you, if you like, and if it turns out good.

. . . . . I am very glad that I misunderstood you about species not having the capacity to vary, though in fact few do give birth to new species. It seems that I am very apt to misunderstand you ; I suppose I am always fancying objections. Your case of the Red Indian shows me that we agree entirely. . . . .

I had a letter yesterday from Thwaites of Ceylon, who was much opposed to me. He now says, "I find that the more familiar I become with your views in connection with the various phenomena of nature, the more they commend themselves to my mind."

* 'Species not Transmutable,' by C. R. Bree, 1860.

C. Darwin to J. M. Rodwell.*

15 Marine Parade, Eastbourne.
November 5th [1860].

MY DEAR SIR,—I am extremely much obliged for your letter, which I can compare only to a plum-pudding, so full it is of good things. I have been rash about the cats : † yet I spoke on what seemed to me, good authority. The Rev. W. D. Fox gave me a list of cases of various foreign breeds in which he had observed the correlation, and for years he had vainly sought an exception. A French paper also gives numerous cases, and one very curious case of a kitten which *gradually* lost the blue colour in its eyes and as gradually acquired its power of hearing. I had not heard of your uncle, Mr. Kirby's case ‡ (whom I, for as long as I can remember, have venerated) of care in breeding cats. I do not know whether Mr. Kirby was your uncle by marriage, but your letters show me that you ought to have Kirby blood in your veins, and that if you had not taken to languages you would have been a first-rate naturalist.

I sincerely hope that you will be able to carry out your intention of writing on the " Birth, Life, and Death of Words." Anyhow, you have a capital title, and some think this the most difficult part of a book. I remember years ago at the Cape of Good Hope, Sir J. Herschell saying to me, I wish some one would treat language as Lyell has treated geology. What a linguist you must be to translate the Koran ! Having a vilely bad head for languages, I feel an awful respect for linguists.

* Rev. J. M. Rodwell, who was at Cambridge with my father, remembers him saying :—" It strikes me that all our knowledge about the structure of our earth is very much like what an old hen would know of a hundred acre field, in a corner of which she is scratching."

† " Cats with blue eyes are invariably deaf," ' Origin of Species,' ed. i. p. 12.

‡ William Kirby, joint author with Spence, of the well-known ' Introduction to Entomology,' 1818.

I do not know whether my brother-in-law, Hensleigh Wedgwood's 'Etymological Dictionary' would be at all in your line ; but he treats briefly on the genesis of words ; and, as it seems to me, very ingeniously. You kindly say that you would communicate any facts which might occur to you, and I am sure that I should be most grateful. Of the multitude of letters which I receive, not one in a thousand is like yours in value.

With my cordial thanks, and apologies for this untidy letter written in haste, pray believe me, my dear Sir,

Yours sincerely obliged,
CH. DARWIN.

C. Darwin to C. Lyell.

November 20th [1860].

. . . . I have not had heart to read Phillips * yet, or a tremendous long hostile review by Professor Bowen in the 4to Mem. of the American Academy of Sciences.† (By the way, I hear Agassiz is going to thunder against me in the next part of the ' Contributions.') Thank you for telling me of the sale of the 'Origin,' of which I had not heard. There will be some time, I presume, a new edition, and I especially want your advice on one point, and you know I think you the wisest of men, and I shall be *absolutely guided by your advice.* It has occurred to me, that it would *perhaps* be a good plan to put a set of notes (some twenty to forty or fifty) to the 'Origin,' which now has none, exclusively devoted to errors of my reviewers. It has occurred to me that where a reviewer has erred, a common reader might err. Secondly, it will show the reader that he must not trust implicitly to reviewers. Thirdly, when any special fact has been attacked, I should like

---

* 'Life on the Earth.'

† "Remarks on the latest form of the Development Theory." By Francis Bowen, Professor of Natural Religion and Moral Philosophy, at Harvard University. 'American Academy of Arts and Sciences,' vol. viii.

to defend it. I would show no sort of anger. I enclose a mere rough specimen, done without any care or accuracy—done from memory alone—to be torn up, just to show the sort of thing that has occurred to me. *Will you do me the great kindness to consider this well?*

It seems to me it would have a good effect, and give some confidence to the reader. It would [be] a horrid bore going through all the reviews.

Yours affectionately,

C. DARWIN.

[Here follow samples of foot-notes, the references to volume and page being left blank. It will be seen that in some cases he seems to have forgotten that he was writing foot-notes, and to have continued as if writing to Lyell :—

* Dr. Bree (p.   ) asserts that I explain the structure of the cells of the Hive Bee by "the exploded doctrine of pressure." But I do not say one word which directly or indirectly can be interpreted into any reference to pressure.

* The 'Edinburgh' Reviewer (vol.   , p.   ) quotes my work as saying that the "dorsal vertebræ of pigeons vary in number, and disputes the fact." I nowhere even allude to the dorsal vertebræ, only to the sacral and caudal vertebræ.

* The 'Edinburgh' Reviewer throws a doubt on these organs being the Branchiæ of Cirripedes. But Professor Owen in 1854 admits, without hesitation, that they are Branchiæ, as did John Hunter long ago.

* The confounded Wealden Calculation to be struck out, and a note to be inserted to the effect that I am convinced of its inaccuracy from a review in the

Saturday Review, and from Phillips, as I see in his Table of Contents that he alludes to it.

* Mr. Hopkins ('Fraser,' vol.   , p.   ) states—I am quoting only from vague memory—that, "I argue in favour of my views from the extreme imperfection of the Geological Record," and says this is the first time in the History of Science he has ever heard of ignorance being adduced as an argument. But I repeatedly admit, in the most emphatic language which I can use, that the imperfect evidence which Geology offers in regard to transitorial forms is most strongly opposed to my views. Surely there is a wide difference in fully admitting an objection, and then in endeavouring to show that it is not so strong as it at first appears, and in Mr. Hopkins's assertion that I found my argument on the Objection.

* I would also put a note to

" Natural Selection," and show how variously it has been misunderstood.

* A writer in the ' Edinburgh Philosophical Journal' denies my statement that the Woodpecker of La Plata never frequents trees. I observed its habits during two years, but, what is more to the purpose, Azara, whose accuracy all admit, is more emphatic than I am in regard to its never frequenting trees. Mr. A. Murray denies that it ought to be called a woodpecker; it has two toes in front and two behind, pointed tail feathers, a long pointed tongue, and the same general form of body, the same manner of flight, colouring and voice. It was classed, until recently, in the same genus—Picus—with all other woodpeckers, but now has been ranked as a distinct genus amongst the Picidæ. It differs from the typical Picus only in the beak, not being quite so strong, and in the upper mandible being slightly arched. I think these facts fully justify my statement that it is "in all essential parts of its organisation" a Woodpecker.]

C. Darwin to T. H. Huxley.

Down, Nov. 22 [1860].

MY DEAR HUXLEY,—For heaven's sake don't write an anti-Darwinian article; you would do it so confoundedly well. I have sometimes amused myself with thinking how I could best pitch into myself, and I believe I could give two or three good digs; but I will see you —— first, before I will try. I shall be very impatient to see the Review.* If it succeeds it may really do much, very much good. . . . .

I heard to-day from Murray that I must set to work at once on a new edition† of the ' Origin.' [Murray] says the Reviews have not improved the sale. I shall always think those early reviews, almost entirely yours, did the subject an *enormous* service. If you have any important suggestions or criticisms to make on any part of the ' Origin,' I should, of course, be very grateful for [them]. For I mean to correct as far as I can, but not enlarge. How you must be wearied with and hate the subject, and it is God's blessing if you do not get to hate me. Adios.

* The first number of the new series of the ' Nat. Hist. Review' appeared in 1861.
† The 3rd edition.

C. Darwin to C. Lyell.

Down, November 24th [1860].

My dear Lyell,—I thank you much for your letter. I had got to take pleasure in thinking how I could best snub my reviewers; but I was determined, in any case, to follow your advice, and, before I had got to the end of your letter, I was convinced of the wisdom of your advice.* What an advantage it is to me to have such friends as you. I shall follow every hint in your letter exactly.

I have just heard from Murray; he says he sold 700 copies at his sale, and that he has not half the number to supply; so that I must begin at once.† . . . .

P.S.—I must tell you one little fact which has pleased me. You may remember that I adduce electrical organs of fish as one of the greatest difficulties which have occurred to me, and —— notices the passage in a singularly disingenuous spirit. Well, McDonnell, of Dublin (a first-rate man), writes to me that he felt the difficulty of the whole case as overwhelming against me. Not only are the fishes which have electric organs very remote in scale, but the organ is near the head in some, and near the tail in others, and supplied by wholly different nerves. It seems impossible that there could be any transition. Some friend, who is much opposed to me, seems to have crowed over McDonnell, who reports that he said to himself, that if Darwin is right, there must be homologous organs both near the head and tail in other non-electric fish.

* "I get on slowly with my new edition. I find that your advice was *excellent*. I can answer all reviews, without any direct notice of them, by a little enlargement here and there, with here and there a new paragraph. Broun alone I shall treat with the respect of giving his objections with his name. I think I shall improve my book a good deal, and add only some twenty pages."— From a letter to Lyell, December 4th, 1860.

† On the third edition of the 'Origin of Species,' published in April 1861.

He set to work, and, by Jove, he has found them!* so that some of the difficulty is removed; and is it not satisfactory that my hypothetical notions should have led to pretty discoveries? McDonnell seems very cautious; he says, years must pass before he will venture to call himself a believer in my doctrine, but that on the subjects which he knows well, viz. Morphology and Embryology, my views accord well, and throw light on the whole subject.

C. Darwin to Asa Gray.

Down, November 26th, 1860.

MY DEAR GRAY,—I have to thank you for two letters. The latter with corrections, written before you received my letter asking for an American reprint, and saying that it was hopeless to print your reviews as a pamphlet, owing to the impossibility of getting pamphlets known. I am very glad to say that the August or second 'Atlantic' article has been reprinted in the 'Annals and Magazine of Natural History'; but I have not yet seen it there. Yesterday I read over with care the third article; and it seems to me, as before, *admirable*. But I grieve to say that I cannot honestly go as far as you do about Design. I am conscious that I am in an utterly hopeless muddle. I cannot think that the world, as we see it, is the result of chance; and yet I cannot look at each separate thing as the result of Design. To take a crucial example, you lead me to infer (p. 414) that you believe "that variation has been led along certain beneficial lines." I cannot believe this; and I think you would have to believe, that the tail of the Fantail was led to vary in the number and direction of its feathers in order to gratify the caprice of a few men. Yet if the Fantail had been a wild bird, and had

* 'On an organ in the Skate, which appears to be the homologue of the electrical organ of the Tor-pedo,' by R. McDonnell, 'Nat. Hist. Review,' 1861, p. 57.

used its abnormal tail for some special end, as to sail before the wind, unlike other birds, every one would have said, "What a beautiful and designed adaptation." Again, I say I am, and shall ever remain, in a hopeless muddle.

Thank you much for Bowen's 4to. review.* The coolness with which he makes all animals to be destitute of reason is simply absurd. It is monstrous at p. 103, that he should argue against the possibility of accumulative variation, and actually leave out, entirely, selection! The chance that an improved Short-horn, or improved Pouter-pigeon, should be produced by accumulative variation without man's selection, is as almost infinity to nothing; so with natural species without natural selection. How capitally in the 'Atlantic' you show that Geology and Astronomy are, according to Bowen, Metaphysics; but he leaves out this in the 4to Memoir.

I have not much to tell you about my Book. I have just heard that Du Bois-Reymond agrees with me. The sale of my book goes on well, and the multitude of reviews has not stopped the sale . . .; so I must begin at once on a new corrected edition. I will send you a copy for the chance of your ever re-reading; but, good Heavens, how sick you must be of it!

C. Darwin to T. H. Huxley.

Down, Dec. 2nd [1860].

. . . . I have got fairly sick of hostile reviews. Nevertheless, they have been of use in showing me when to expatiate a little and to introduce a few new discussions. *Of course* I will send you a copy of the new edition

I entirely agree with you, that the difficulties on my notions are terrific, yet having seen what all the Reviews have said against me, I have far more confidence in the *general* truth of the doctrine than I formerly had. Another thing

* 'Memoirs of the American Academy of Arts and Sciences,' vol. viii.

gives me confidence, viz. that some who went half an inch with me now go further, and some who were bitterly opposed are now less bitterly opposed. And this makes me feel a little disappointed that you are not inclined to think the general view in some slight degree more probable than you did at first. This I consider rather ominous. Otherwise I should be more contented with your degree of belief. I can pretty plainly see that, if my view is ever to be generally adopted, it will be by young men growing up and replacing the old workers, and then young ones finding that they can group facts and search out new lines of investigation better on the notion of descent, than on that of creation. But forgive me for running on so egotistically. Living so solitary as I do, one gets to think in a silly manner of one's own work.

Ever yours very sincerely,
C. DARWIN.

C. Darwin to J. D. Hooker

Down, December 11th [1860].

. . . . . . I heard from A. Gray this morning; at my suggestion he is going to reprint the three 'Atlantic' articles as a pamphlet, and send 250 copies to England, for which I intend to pay half the cost of the whole edition, and shall give away, and try to sell by getting a few advertisements put in, and if possible notices in Periodicals.

. . . . . . David Forbes has been carefully working the Geology of Chile, and as I value praise for accurate observation far higher than for any other quality, forgive (if you can) the *insufferable* vanity of my copying the last sentence in his note: " I regard your Monograph on Chile as, without exception, one of the finest specimens of Geological enquiry." I feel inclined to strut like a Turkey-cock !

CHAPTER VIII.

THE SPREAD OF EVOLUTION.

1861–1862.

[THE beginning of the year 1861 saw my father with the third chapter of 'The Variation of Animals and Plants' still on his hands. It had been begun in the previous August, and was not finished until March 1861. He was, however, for part of this time (I believe during December 1860 and January 1861) engaged in a new edition (3000 copies) of the 'Origin,' which was largely corrected and added to, and was published in April 1861.

With regard to this, the third edition, he wrote to Mr. Murray in December 1860 :—

"I shall be glad to hear when you have decided how many copies you will print off—the more the better for me in all ways, as far as compatible with safety ; for I hope never again to make so many corrections, or rather additions, which I have made in hopes of making my many rather stupid reviewers at least understand what is meant. I hope and think I shall improve the book considerably."

An interesting feature in the new edition was the "Historical Sketch of the Recent Progress of Opinion on the Origin of Species " * which now appeared for the first time, and was continued in the later editions of the work. It bears a strong

---

* The Historical Sketch had already appeared in the first German edition (1860) and the American edition. Bronn states in the German edition (footnote, p. 1) that it was his critique in the ' N. Jahrbuch für Mineralogie ' that suggested the idea of such a sketch to my father.

impress of the author's personal character in the obvious wish to do full justice to all his predecessors,—though even in this respect it has not escaped some adverse criticism.

Towards the end of the present year (1861), the final arrangements for the first French edition of the 'Origin' were completed, and in September a copy of the third English edition was despatched to Mdlle. Clémence Royer, who undertook the work of translation. The book was now spreading on the Continent, a Dutch edition had appeared, and, as we have seen, a German translation had been published in 1860. In a letter to Mr. Murray (September 10, 1861), he wrote, " My book seems exciting much attention in Germany, judging from the number of discussions sent me." The silence had been broken, and in a few years the voice of German science was to become one of the strongest of the advocates of evolution.

During all the early part of the year (1861) he was working at the mass of details which are marshalled in order in the early chapters of 'Animals and Plants.' Thus in his Diary occur the laconic entries, " May 16, Finished Fowls (eight weeks); May 31, Ducks."

On July 1, he started, with his family, for Torquay, where he remained until August 27—a holiday which he characteristically enters in his diary as "eight weeks and a day." The house he occupied was in Hesketh Crescent, a pleasantly placed row of houses close above the sea, somewhat removed from what was then the main body of the town, and not far from the beautiful cliffed coast-line in the neighbourhood of Anstey's Cove.

During the Torquay holiday, and for the remainder of the year, he worked at the fertilisation of orchids. This part of the year 1861 is not dealt with in the present chapter, because (as explained in the preface) the record of his life, as told in his letters, seems to become clearer when the whole of his botanical work is placed together and treated separately.

The present series of chapters will, therefore, include only the progress of his works in the direction of a general amplification of the 'Origin of Species'—e.g., the publication of 'Animals and Plants,' 'Descent of Man,' &c.]

C. Darwin to J. D. Hooker.

Down, Jan. 15 [1861].

MY DEAR HOOKER,—The sight of your handwriting always rejoices the very cockles of my heart. . . . .

I most fully agree to what you say about Huxley's Article,* and the power of writing. . . . . The whole review seems to me excellent. How capitally Oliver has done the résumé of botanical books. Good Heavens, how he must have read ! . . . .

I quite agree that Phillips † is unreadably dull. You need not attempt Bree. ‡ . . . .

* 'Natural History Review,' 1861, p. 67, "On the Zoological Relations of Man with the Lower Animals." This memoir had its origin in a discussion at the previous meeting of the British Association, when Professor Huxley felt himself " compelled to give a diametrical contradiction to certain assertions respecting the differences which obtain between the brains of the higher apes and of man, which fell from Professor Owen." But in order that his criticisms might refer to deliberately recorded words, he bases them on Professor Owen's paper, " On the Characters, &c., of the Class Mammalia," read before the Linnean Society in February and April, 1857, in which he proposed to place man not only in a distinct order, but in "a distinct sub-class of the Mammalia "—the Archencephala.

† 'Life on the Earth' (1860), by Prof. Phillips, containing the substance of the Rede Lecture (May 1860).

‡ The following sentence (p. 16) from ' Species not Transmutable,' by Dr. Bree, illustrates the degree in which he understood the ' Origin of Species ': " The only real difference between Mr. Darwin and his two predecessors " [Lamarck and the ' Vestiges '] " is this :—that while the latter have each given a mode by which they conceive the great changes they believe in have been brought about, Mr. Darwin does no such thing." After this we need not be surprised at a passage in the preface : " No one has derived greater pleasure than I have in past days from the study of Mr. Darwin's other works, and no one has felt a greater degree of regret that he should have imperilled his fame by the publication of his treatise upon the ' Origin of Species.'"

If you come across Dr. Freke on the 'Origin of Species by means of Organic Affinity,' read a page here and there. . . . He tells the reader to observe [that his result] has been arrived at by "induction," whereas all my results are arrived at only by "analogy." I see a Mr. Neale has read a paper before the Zoological Society on 'Typical Selection;' what it means I know not. I have not read H. Spencer, for I find that I must more and more husband the very little strength which I have. I sometimes suspect I shall soon entirely fail. . . . . As soon as this dreadful weather gets a little milder, I must try a little water cure. Have you read the 'Woman in White'? the plot is wonderfully interesting. I can recommend a book which has interested me greatly, viz. Olmsted's 'Journey in the Back Country.' It is an admirably lively picture of man and slavery in the Southern States. . . . .

C. Darwin to C. Lyell.

February 2, 1861.

MY DEAR LYELL,—I have thought you would like to read the enclosed passage in a letter from A. Gray (who is printing his reviews as a pamphlet,* and will send copies to England), as I think his account is really favourable in a high degree to us :—

"I wish I had time to write you an account of the lengths to which Bowen and Agassiz, each in their own way, are going. The first denying all heredity (all transmission except specific) whatever. The second coming near to deny that we are genetically descended from our great-great-grandfathers; and insisting that evidently affiliated languages, e.g. Latin, Greek, Sanscrit, owe none of their similarities to a community of origin, are all autochthonal; Agassiz admits that

---

* "Natural Selection not inconsistent with Natural Theology," from the 'Atlantic Monthly' for July, August, and October, 1860; published by Trübner.

the derivation of languages, and that of species or forms, stand on the same foundation, and that he must allow the latter if he allows the former, which I tell him is perfectly logical."

Is not this marvellous?

> Ever yours,
> C. DARWIN.

C. Darwin to J. D. Hooker.

> Down, Feb. 4 [1861].

MY DEAR HOOKER,—I was delighted to get your long chatty letter, and to hear that you are thawing towards science. I almost wish you had remained frozen rather longer; but do not thaw too quickly and strongly. No one can work long as you used to do. Be idle; but I am a pretty man to preach, for I cannot be idle, much as I wish it, and am never comfortable except when at work. The word holiday is written in a dead language for me, and much I grieve at it. We thank you sincerely for your kind sympathy about poor H. [his daughter]. . . . . She has now come up to her old point, and can sometimes get up for an hour or two twice a day . . . . Never to look to the future or as little as possible is becoming our rule of life. What a different thing life was in youth with no dread in the future; all golden, if baseless, hopes.

. . . . With respect to the 'Natural History Review' I can hardly think that ladies would be so very sensitive about "lizards' guts;" but the publication is at present certainly a sort of hybrid, and original illustrated papers ought hardly to appear in a review. I doubt its ever paying; but I shall much regret if it dies. All that you say seems very sensible, but could a review in the strict sense of the word be filled with readable matter?

I have been doing little, except finishing the new edition

of the 'Origin,' and crawling on most slowly with my
volume of 'Variation under Domestication.' . . . .

[The following letter refers to Mr. Bates's paper, "Contributions to an Insect Fauna of the Amazon Valley," in the
'Transactions' of the Entomological Society.' vol. 5, N.S.*
Mr. Bates points out that with the return, after the glacial
period, of a warmer climate in the equatorial regions, the
"species then living near the equator would retreat north
and south to their former homes, leaving some of their con-
geners, slowly modified subsequently . . . to re-people the zone
they had forsaken." In this case the species now living at
the equator ought to show clear relationship to the species
inhabiting the regions about the 25th parallel, whose distant
relatives they would of course be. But this is not the case,
and this is the difficulty my father refers to. Mr. Belt has
offered an explanation in his 'Naturalist in Nicaragua'
(1874), p. 266. "I believe the answer is that there was much
extermination during the glacial period, that many species
(and some genera, &c., as, for instance, the American horse),
did not survive it . . . . but that a refuge was found for
many species on lands now below the ocean, that were
uncovered by the lowering of the sea, caused by the immense
quantity of water that was locked up in frozen masses on the
land."]

C. Darwin to J. D. Hooker.

Down, 27th [March 1861].

MY DEAR HOOKER,—I had intended to have sent you
Bates's article this very day. I am so glad you like it. I have
been extremely much struck with it. How well he argues,
and with what crushing force against the glacial doctrine.
I cannot wriggle out of it: I am dumbfounded ; yet I do
believe that some explanation some day will appear, and I

* The paper was read Nov. 24, 1860.

cannot give up equatorial cooling. It explains so much and harmonises with so much. When you write (and much interested I shall be in your letter) please say how far floras are generally uniform in generic character from 0° to 25° N. and S.

Before reading Bates, I had become thoroughly dissatisfied with what I wrote to you. I hope you may get Bates to write in the 'Linnean.'

Here is a good joke : H. C. Watson (who, I fancy and hope, is going to review the new edition * of the 'Origin') says that in the first four paragraphs of the introduction, the words "I," "me," "my," occur forty-three times! I was dimly conscious of the accursed fact. He says it can be explained phrenologically, which I suppose civilly means, that I am the most egotistically self-sufficient man alive; perhaps so. I wonder whether he will print this pleasing fact; it beats hollow the parentheses in Wollaston's writing.

I am, *my* dear Hooker, ever yours,

C. DARWIN.

P.S.—Do not spread this pleasing joke; it is rather too biting.

C. Darwin to J. D. Hooker.

Down, [April] 23 ? [1861.]

. . . . I quite agree with what you say on Lieutenant Hutton's Review † (who he is I know not); it struck me as very original. He is one of the very few who see that the change of species cannot be directly proved, and that the doctrine must sink or swim according as it groups and explains phenomena. It is really curious how few judge it in this way, which is clearly the right way. I have been much

* Third edition of 2000 copies, published in April 1861.

† In the 'Geologist,' 1861, p. 132, by Lieutenant Frederick Wollaston Hutton, of the Staff College. The 'Geologist' was afterwards merged in the 'Geological Magazine.'

interested by Bentham's paper* in the N. H. R., but it would not, of course, from familiarity, strike you as it did me. I liked the whole; all the facts on the nature of close and varying species. Good Heavens! to think of the British botanists turning up their noses, and saying that he knows nothing of British plants! I was also pleased at his remarks on classification, because it showed me that I wrote truly on this subject in the 'Origin.' I saw Bentham at the Linnean Society, and had some talk with him and Lubbock, and Edgeworth, Wallich, and several others. I asked Bentham to give us his ideas of species; whether partially with us or dead against us, he would write *excellent* matter. He made no answer, but his manner made me think he might do so if urged; so do you attack him. Every one was speaking with affection and anxiety of Henslow.† I dined with Bell at the Linnean Club, and liked my dinner. . . . . Dining out is such a novelty to me that I enjoyed it. Bell has a real good heart. I liked Rolleston's paper, but I never read anything so obscure and not self-evident as his 'Canons.'‡ . . . . I called on R. Chambers, at his very nice house in St. John's Wood, and had a very pleasant half-hour's talk; he is really a capital fellow. He made one good remark and chuckled over it, that the laymen universally had treated the controversy on the 'Essays and Reviews' as a merely professional subject, and had not joined in it, but had left it to the clergy. I shall be anxious for your next letter about Henslow.§ Farewell, with sincere sympathy, my old friend,

C. DARWIN.

---

* "On the Species and Genera of Plants, &c.," 'Natural History Review,' 1861, p. 133.

† Prof. Henslow was in his last illness.

‡ George Rolleston, M.D., F.R.S., b. 1829, d. 1881. Linacre Professor of Anatomy and Physiology at Oxford. A man of much learning who left but few published works, among which may be mentioned his handbook, 'Forms of Animal Life.' For the 'Canons,' see 'Nat. Hist. Review,' 1861, p. 206.

§ Sir Joseph Hooker was Prof. Henslow's son-in-law.

P.S.—We are very much obliged for the 'London Review.' We like reading much of it, and the science is incomparably better than in the *Athenæum*. You shall not go on very long sending it, as you will be ruined by pennies and trouble, but I am under a horrid spell to the *Athenæum* and the *Gardeners' Chronicle*, but I have taken them in for so many years, that I *cannot* give them up.

[The next letter refers to Lyell's visit to the Biddenham gravel-pits near Bedford in April 1861. The visit was made at the invitation of Mr. James Wyatt, who had recently discovered two stone implements " at the depth of thirteen feet from the surface of the soil," resting " immediately on solid beds of oolitic-limestone." *   Here, says Sir C. Lyell, " I . . . . for the first time, saw evidence which satisfied me of the chronological relations of those three phenomena—the antique tools, the extinct mammalia, and the glacial formation."]

C. Darwin to C. Lyell.

Down, April 12 [1861].

MY DEAR LYELL,—I have been most deeply interested by your letter. You seem to have done the grandest work, and made the greatest step, of any one with respect to man.

It is an especial relief to hear that you think the French superficial deposits are deltoid and semi-marine ; but two days ago I was saying to a friend, that the unknown manner of the accumulation of these deposits, seemed the great blot in all the work done. I could not stomach debacles or lacustrine beds. It is grand. I remember Falconer told me that he

---

* 'Antiquity of Man,' fourth edition, p. 214.

thought some of the remains in the Devonshire caverns were pre-glacial, and this, I presume, is now your conclusion for the older celts with hyena and hippopotamus. It is grand. What a fine long pedigree you have given the human race!

I am sure I never thought of parallel roads having been accumulated during subsidence. I think I see some difficulties on this view, though, at first reading your note, I jumped at the idea. But I will think over all I saw there. I am (stomacho volente) coming up to London on Tuesday to work on cocks and hens, and on Wednesday morning, about a quarter before ten, I will call on you (unless I hear to the contrary), for I long to see you. I congratulate you on your grand work.

Ever yours,
C. DARWIN.

P.S.—Tell Lady Lyell that I was unable to digest the funereal ceremonies of the ants, notwithstanding that Erasmus has often told me that I should find some day that they have their bishops. After a battle I have always seen the ants carry away the dead for food. Ants display the utmost economy, and always carry away a dead fellow-creature as food. But I have just forwarded two most extraordinary letters to Busk, from a backwoodsman in Texas, who has evidently watched ants carefully, and declares most positively that they plant and cultivate a kind of grass for store food, and plant other bushes for shelter! I do not know what to think, except that the old gentleman is not fibbing intentionally. I have left the responsibility with Busk whether or no to read the letters.*

* *I.e.* to read them before the Linnean Society.

C. Darwin to Thomas Davidson.*

Down, April 26, 1861.

MY DEAR SIR,—I hope that you will excuse me for venturing to make a suggestion to you which I am perfectly well aware it is a very remote chance that you would adopt. I do not know whether you have read my 'Origin of Species'; in that book I have made the remark, which I apprehend will be universally admitted, that˙ *as a whole*, the fauna of any formation is intermediate in character between that of the formations above and below. But several really good judges have remarked to me how desirable it would be that this should be exemplified and worked out in some detail and with some single group of beings. Now every one will admit that no one in the world could do this better than you with Brachiopods. The result might turn out very unfavourable to the views which I hold; if so, so much the better for those who are opposed to me.† But I am inclined to suspect that on the whole it would be favourable to the notion of descent with modification; for about a year ago, Mr. Salter ‡ in the museum in Jermyn Street, glued on a board some

* Thomas Davidson, F.R.S., born in Edinburgh, May 17, 1817; died 1885. His researches were chiefly connected with the sciences of geology and palæontology, and were directed especially to the elucidation of the characters, classification, history, geological and geographical distribution of recent and fossil Brachiopoda. On this subject he brought out an important work, 'British Fossil Brachiopoda,' 5 vols. 4to. (Cooper, 'Men of the Time,' 1884.)

† "Mr. Davidson is not at all a full believer in great changes of species, which will make his work all the more valuable."—C. Dar-

win to R. Chambers (April 30, 1861).

‡ John William Salter; b. 1820, d. 1869. He entered the service of the Geological Survey in 1846, and ultimately became its Palæontologist, on the retirement of Edward Forbes, and gave up the office in 1863. He was associated with several well-known naturalists in their work—with Sedgwick, Murchison, Lyell, Ramsay, and Huxley. There are sixty entries under his name in the Royal Society Catalogue. The above facts are taken from an obituary notice of Mr. Salter in the 'Geological Magazine,' 1869.

Spirifers, &c., from three palæozoic stages, and arranged them
in single and branching lines, with horizontal lines marking
the formations (like the diagram in my book, if you know
it), and the result seemed to me very striking, though I was
too ignorant fully to appreciate the lines of affinities.   I
longed to have had these shells engraved, as arranged by
Mr. Salter, and connected by dotted lines, and would have
gladly paid the expense : but I could not persuade Mr. Salter
to publish a little paper on the subject.   I can hardly doubt
that many curious points would occur to any one thoroughly
instructed in the subject, who would consider a group of
beings under this point of view of descent with modification.
All those forms which have come down from an ancient
period very slightly modified ought, I think, to be omitted,
and those forms alone considered which have undergone
considerable change at each successive epoch.   My fear is
whether brachiopods have changed enough.   The absolute
amount of difference of the forms in such groups at the
opposite extremes of time ought to be considered, and how
far the early forms are intermediate in character between
those which appeared much later in time.   The antiquity of
a group is not really diminished, as some seem vaguely to
think, because it has transmitted to the present day closely
allied forms.   Another point is how far the succession of each
genus is unbroken, from the first time it appeared to its
extinction, with due allowance made for formations poor in
fossils.   I cannot but think that an important essay (far more
important than a hundred literary reviews) might be written
by one like yourself, and without very great labour.   I know
it is highly probable that you may not have leisure, or not
care for, or dislike the subject, but I trust to your kindness
to forgive me for making this suggestion.   If by any extraordinary good fortune you were inclined to take up this
notion, I would ask you to read my Chapter X. on Geological Succession.   And I should like in this case to be

permitted to send you a copy of the new edition, just published, in which I have added and corrected somewhat in
Chapters IX. and X.

Pray excuse this long letter, and believe me,

My dear Sir, yours very faithfully,

C. DARWIN.

P.S.—I write so bad a hand that I have had this note
copied.

C. Darwin to Thomas Davidson.

Down, April 30, 1861.

MY DEAR SIR,—I thank you warmly for your letter ; I did
not in the least know that you had attended to my work.   I
assure you that the attention which you have paid to it, considering your knowledge and the philosophical tone of your
mind (for I well remember one remarkable letter you wrote
to me, and have looked through your various publications),
I consider one of the highest, perhaps the very highest, compliments which I have received.   I live so solitary a life that
I do not often hear what goes on, and I should much like to
know in what work you have published some remarks on my
book.   I take a deep interest in the subject, and I hope not
simply an egotistical interest ; therefore you may believe how
much your letter has gratified me ; I am perfectly contented
if any one will fairly consider the subject, whether or not he
fully or only very slightly agrees with me.   Pray do not
think that I feel the least surprise at your demurring to a
ready acceptance ; in fact, I should not much respect anyone's
judgment who did so : that is, if I may judge others from
the long time which it has taken me to go round.   Each
stage of belief cost me years.   The difficulties are, as you say,
many and very great ; but the more I reflect, the more they
seem to me to be due to our underestimating our ignorance.
I belong so much to old times that I find that I weigh

the difficulties from the imperfection of the geological record, heavier than some of the younger men. I find, to my astonishment and joy, that such good men as Ramsay, Jukes, Geikie, and one old worker, Lyell, do not think that I have in the least exaggerated the imperfection of the record.* If my views ever are proved true, our current geological views will have to be considerably modified. My greatest trouble is, not being able to weigh the direct effects of the long-continued action of changed conditions of life without any selection, with the action of selection on mere accidental (so to speak) variability. I oscillate much on this head, but generally return to my belief that the direct action of the conditions of life have not been great. At least this direct action can have played an extremely small part in producing all the numberless and beautiful adaptations in every living creature. With respect to a person's belief, what does rather surprise me is that any one (like Carpenter) should be willing *to go so very far* as to believe that all birds may have descended from one parent, and not go a little farther and include all the members of the same great division ; for on such a scale of belief, all the facts in Morphology and in Embryology (the most important in my opinion of all subjects) become mere Divine mockeries. . . . . I cannot express how profoundly glad I am that some day you will publish your theoretical view on the modification and endurance of

---

* Professor Sedgwick treated this part of the 'Origin of Species' very differently, as might have been expected from his vehement objection to Evolution in general. In the article in the *Spectator* of March 24, 1860, already noticed, Sedgwick wrote : "We know the complicated organic phenomena of the Mesozoic (or Oolitic) period. It defies the transmutationist at every step. Oh ! but the document, says Darwin, is a fragment ; I will interpolate long periods to account for all the changes. I say, in reply, if you deny my conclusion, grounded on positive evidence, I toss back your conclusion, derived from negative evidence,—the inflated cushion on which you try to bolster up the defects of your hypothesis." [The punctuation of the imaginary dialogue is slightly altered from the original, which is obscure in one place.]

Brachiopodous species ; I am sure it will be a most valuable
contribution to knowledge.

Pray forgive this very egotistical letter, but you yourself
are partly to blame for having pleased me so much. I have
told Murray to send a copy of my new edition to you, and
have written your name.

With cordial thanks, pray believe me, my dear Sir,.

Yours very sincerely,

CH. DARWIN.

[In Mr. Davidson's Monograph on British Brachiopoda,.
published shortly afterwards by the Palæontographical Society,
results such as my father anticipated were to some extent
obtained. "No less than fifteen commonly received species.
are demonstrated by Mr. Davidson by the aid of a long series
of transitional forms to appertain to . . . one type." *

In the autumn of 1860, and the early part of 1861, my
father had a good deal of correspondence with Professor
Asa Gray on a subject to which reference has already been
made—the publication, in the form of a pamphlet, of Professor Gray's three articles in the July, August, and October
numbers of the 'Atlantic Monthly,' 1860. The pamphlet was
published by Messrs. Trübner, with reference to whom my
father wrote, "Messrs. Trübner have been most liberal and
kind, and say they shall make no charge for all their trouble.
I have settled about a few advertisements, and they will
gratuitously insert one in their own periodicals."

The reader will find these articles republished in Dr. Gray's
'Darwiniana,' p. 87, under the title "Natural Selection not
inconsistent with Natural Theology." The pamphlet found
many admirers among those most capable of judging of its
merits, and my father believed that it was of much value in
lessening opposition, and making converts to Evolution. His.

* Lyell, 'Antiquity of Man,' first edition, p. 428.

high opinion of it is shown not only in his letters, but by the fact that he inserted a special notice of it in a most prominent place in the third edition of the 'Origin.' Lyell, among others, recognised its value as an antidote to the kind of criticism from which the cause of Evolution suffered. Thus my father wrote to Dr. Gray :—"Just to exemplify the use of your pamphlet, the Bishop of London was asking Lyell what he thought of the review in the 'Quarterly,' and Lyell answered, 'Read Asa Gray in the 'Atlantic.'" It comes out very clearly that in the case of such publications as Dr. Gray's, my father did not rejoice over the success of his special view of Evolution, viz. that modification is mainly due to Natural Selection ; on the contrary, he felt strongly that the really important point was that the doctrine of Descent should be accepted. Thus he wrote to Professor Gray (May 11, 1863), with reference to Lyell's 'Antiquity of Man' :—

" You speak of Lyell as a judge ; now what I complain of is that he declines to be a judge . . . . I have sometimes almost wished that Lyell had pronounced against me. When I say ' me,' I only mean *change of species by descent.* That seems to me the turning-point. Personally, of course, I care much about Natural Selection ; but that seems to me utterly unimportant, compared to the question of Creation *or* Modification."]

C. Darwin to Asa Gray.

Down, April 11 [1861].

MY DEAR GRAY,—I was very glad to get your photograph : I am expecting mine, which I will send off as soon as it comes. It is an ugly affair, and I fear the fault does not lie with the photographer. . . . . Since writing last, I have had several letters full of the highest commendation of your Essay ; all agree that it is by far the best thing written, and I do not doubt it has done the 'Origin' much good. I have not yet heard how it has sold. You will have seen a review in the

Gardeners' Chronicle. Poor dear Henslow, to whom I owe much, is dying, and Hooker is with him. Many thanks for two sets of sheets of your Proceedings. I cannot understand what Agassiz is driving at. You once spoke, I think, of Professor Bowen as a very clever man. I should have thought him a singularly unobservant man from his writings. He never can have seen much of animals, or he would have seen the difference of old and wise dogs and young ones. His paper about hereditariness beats everything. Tell a breeder that he might pick out his worst *individual* animals and breed from them, and hope to win a prize, and he would think you . . . insane.

[Professor Henslow died on May 16, 1861, from a complication of bronchitis, congestion of the lungs, and enlargement of the heart. His strong constitution was slow in giving way, and he lingered for weeks in a painful condition of weakness, knowing that his end was near, and looking at death with fearless eyes. In Mr. Blomefield's (Jenyns) 'Memoir of Henslow' (1862) is a dignified and touching description of Prof. Sedgwick's farewell visit to his old friend. Sedgwick said afterwards that he had never seen "a human being whose soul was nearer heaven."

My father wrote to Sir J. D. Hooker on hearing of Henslow's death, "I fully believe a better man never walked this earth."

He gave his impressions of Henslow's character in Mr. Blomefield's 'Memoir.' In reference to these recollections he wrote to Sir J. D. Hooker (May 30, 1861):—

" This morning I wrote my recollections and impressions of character of poor dear Henslow about the year 1830. I liked the job, and so have written four or five pages, now being copied. I do not suppose you will use all, of course you can chop and change as much as you like. If more than a sentence is used, I should like to see a proof-page, as I never can write decently till I see it in print. Very likely some of my remarks may appear too trifling, but I thought it best to give my thoughts as they arose, for you or Jenyns to use as you think fit.

"You will see that I have exceeded your request, but, as I said when I began, I took pleasure in writing my impression of his admirable character."]

C. Darwin to Asa Gray.

Down, June 5 [1861].

MY DEAR GRAY,—I have been rather extra busy, so have been slack in answering your note of May 6th. I hope you have received long ago the third edition of the 'Origin.' . . . . I have heard nothing from Trübner of the sale of your Essay, hence fear it has not been great; I wrote to say you could supply more. I sent a copy to Sir J. Herschel, and in his new edition of his 'Physical Geography' he has a note on the 'Origin of Species,' and agrees, to a certain limited extent, but puts in a caution on design—much like yours. . . . . I have been led to think more on this subject of late, and grieve to say that I come to differ more from you. It is not that designed variation makes, as it seems to me, my deity "Natural Selection" superfluous, but rather from studying, lately, domestic variation, and seeing what an enormous field of undesigned variability there is ready for natural selection to appropriate for any purpose useful to each creature.

I thank you much for sending me your review of Phillips.* I remember once telling you a lot of trades which you ought to have followed, but now I am convinced that you are a born reviewer. By Jove, how well and often you hit the nail on the head! You rank Phillips's book higher than I do, or than Lyell does, who thinks it fearfully retrograde. I amused myself by parodying Phillips's argument as applied to domestic variation; and you might thus prove that the duck or

* 'Life on the Earth,' 1860.

pigeon has not varied because the goose has not, though more anciently domesticated, and no good reason can be assigned why it has not produced many varieties. . . . .

I never knew the newspapers so profoundly interesting. North America does not do England justice; I have not seen or heard of a soul who is not with the North. Some few, and I am one of them, even wish to God, though at the loss of millions of lives, that the North would proclaim a crusade against slavery. In the long-run, a million horrid deaths would be amply repaid in the cause of humanity. What wonderful times we live in! Massachusetts seems to show noble enthusiasm. Great God! how I should like to see the greatest curse on earth—slavery—abolished!

Farewell. Hooker has been absorbed with poor dear revered Henslow's affairs. Farewell.

Ever yours,
C. DARWIN.

Hugh Falconer to C. Darwin.

31 Sackville St., W., June 23, 1861.

MY DEAR DARWIN.—I have been to Adelsberg cave and brought back with me a live *Proteus anguinus*, designed for you from the moment I got it; *i.e.* if you have got an aquarium and would care to have it. I only returned last night from the Continent, and hearing from your brother that you are about to go to Torquay, I lose no time in making you the offer. The poor dear animal is still alive—although it has had no appreciable means of sustenance for a month— and I am most anxious to get rid of the responsibility of starving it longer. In your hands it will thrive and have a fair chance of being developed without delay into some type of the Columbidæ—say a Pouter or a Tumbler.

My dear Darwin, I have been rambling through the north of Italy, and Germany lately. Everywhere have I heard your views and your admirable essay canvassed—the views of course often dissented from, according to the special bias of the speaker—but the work, its honesty of purpose, grandeur of conception, felicity of illustration, and courageous exposition, always referred to in terms of the highest admiration. And among your warmest friends no one rejoiced more heartily in the just appreciation of Charles Darwin than did,

Yours very truly,

H. FALCONER.

C. Darwin to Hugh Falconer.

Down [June 24, 1861].

MY DEAR FALCONER.—I have just received your note, and by good luck a day earlier than properly, and I lose not a moment in answering you, and thanking you heartily for your offer of the valuable specimen ; but I have no aquarium and shall soon start for Torquay, so that it would be a thousand pities that I should have it. Yet I should certainly much like to see it, but I fear it is impossible. Would not the Zoological Society be the best place? and then the interest which many would take in this extraordinary animal would repay you for your trouble.

Kind as you have been in taking this trouble and offering me this specimen, to tell the truth I value your note more than the specimen. I shall keep your note amongst a very few precious letters. Your kindness has quite touched me.

Yours affectionately and gratefully,

CH. DARWIN.

C. Darwin to J. D. Hooker.

2 Hesketh Crescent, Torquay,
July 13 [1861].

. . . I hope Harvey is better ; I got his review * of me a day or two ago, from which I infer he must be convalescent ;

* The 'Dublin Hospital Gazette,' May 15, 1861. The passage referred to is at p. 150.

it's very good and fair ; but it is funny to see a man argue on the succession of animals from Noah's Deluge ; as God did not then wholly destroy man, probably he did not wholly destroy the races of other animals at each geological period ! I never expected to have a helping hand from the Old Testament. . . .

C. Darwin to C. Lyell.

2, Hesketh Crescent, Torquay,
July 20 [1861].

MY DEAR LYELL.—I sent you two or three days ago a duplicate of a good review of the 'Origin' by a Mr. Maw,* evidently a thoughtful man, as I thought you might like to have it, as you have so many. . . .

This is a quite charming place, and I have actually walked, I believe, good two miles out and back, which is a grand feat.

I saw Mr. Pengelly † the other day, and was pleased at his enthusiasm. I do not in the least know whether you are in London. Your illness must have lost you much time, but I hope you have nearly got your great job of the new edition finished. You must be very busy, if in London, so I will be generous, and on honour bright do not expect any answer to this dull little note. . . .

C. Darwin to Asa Gray.

Down, September 17 [1861 ?]

MY DEAR GRAY. —I thank you sincerely for your very long and interesting letter, political and scientific, of August 27th

---

* Mr. George Maw, of Benthall Hall. The review was published in the 'Zoologist,' July, 1861. On the back of my father's copy is written, "Must be consulted before new edit. of 'Origin'"—words which are wanting on many more pretentious notices, on which frequently occur my father's brief 0/-, or " nothing new."

† William Pengelly, the geologist, and well-known explorer of the Devonshire caves.

and 29th, and Sept. 2nd received this morning. I agree with much of what you say, and I hope to God we English are utterly wrong in doubting (1) whether the N. can conquer the S. ; (2) whether the N. has many friends in the South, and (3) whether you noble men of Massachusetts are right in transferring your own good feelings to the men of Washington. Again I say I hope to God we are wrong in doubting on these points. It is number (3) which alone causes England not to be enthusiastic with you. What it may be in Lancashire I know not, but in S. England cotton has nothing whatever to do with our doubts. If abolition does follow with your victory, the whole world will look brighter in my eyes, and in many eyes. It would be a great gain even to stop the spread of slavery into the Territories ; if that be possible without abolition, which I should have doubted. You ought not to wonder so much at England's coldness, when you recollect at the commencement of the war how many propositions were made to get things back to the old state with the old line of latitude. But enough of this, all I can say is that Massachusetts and the adjoining States have the full sympathy of every good man whom I see ; and this sympathy would be extended to the whole Federal States, if we could be persuaded that your feelings were at all common to them. But enough of this. It is out of my line, though I read every word of news, and formerly well studied Olmsted. . . . .

Your question what would convince me of Design is a poser. If I saw an angel come down to teach us good, and I was convinced from others seeing him that I was not mad, I should believe in design. If I could be convinced thoroughly that life and mind was in an unknown way a function of other imponderable force, I should be convinced. If man was made of brass or iron and no way connected with any other organism which had ever lived, I should perhaps be convinced. But this is childish writing.

I have lately been corresponding with Lyell, who, I think, adopts your idea of the stream of variation having been led or designed.   I have asked him (and he says he will hereafter reflect and answer me) whether he believes that the shape of my nose was designed.   If he does I have nothing more to say.   If not, seeing what Fanciers have done by selecting individual differences in the nasal bones of pigeons, I must think that it is illogical to suppose that the variations, which natural selection preserves for the good of any being, have been designed.   But I know that I am in the same sort of muddle (as I have said before) as all the world seems to be in with respect to free will, yet with everything supposed to have been foreseen or pre-ordained.

Farewell, my dear Gray, with many thanks for your interesting letter.

Your unmerciful correspondent,

C. DARWIN.

C. Darwin to H. W. Bates.

Down, Dec. 3 [1861].

MY DEAR SIR.—I thank you for your extremely interesting letter, and valuable references, though God knows when I shall come again to this part of my subject.   One cannot of course judge of style when one merely hears a paper,* but yours seemed to me very clear and good.   Believe me that I estimate its value most highly.   Under a general point of view, I am quite convinced (Hooker and Huxley took the same view some months ago) that a philosophic view of nature can solely be driven into naturalists by treating special subjects as you have done.   Under a special point of view, I think you have solved one of the most perplexing problems which could be given to solve.   I am glad to hear from Hooker

* On Mimetic Butterflies, read before the Linnean Soc., Nov. 21, 1861.   For my father's opinion of it when published, see p. 391.

that the Linnean Society will give plates if you can get drawings. . . .

Do not complain of want of advice during your travels; I dare say part of your great originality of views may be due to the necessity of self-exertion of thought. I can understand that your reception at the British Museum would damp you; they are a very good set of men, but not the sort to appreciate your work. In fact I have long thought that *too much* systematic work [and] description somehow blunts the faculties. The general public appreciates a good dose of reasoning, or generalisation, with new and curious remarks on habits, final causes, &c. &c., far more than do the regular naturalists.

I am extremely glad to hear that you have begun your travels . . . I am very busy, but I shall be *truly* glad to render any aid which I can by reading your first chapter or two. I do not think I shall be able to correct style, for this reason, that after repeated trials I find I cannot correct my own style till I see the MS. in type. Some are born with a power of good writing, like Wallace; others like myself and Lyell have to labour very hard and slowly at every sentence. I find it a very good plan, when I cannot get a difficult discussion to please me, to fancy that some one comes into the room and asks me what I am doing; and then try at once and explain to the imaginary person what it is all about. I have done this for one paragraph to myself several times, and sometimes to Mrs. Darwin, till I see how the subject ought to go. It is, I think, good to read one's MS. aloud. But style to me is a great difficulty; yet some good judges think I have succeeded, and I say this to encourage you.

What *I think* I can do will be to tell you whether parts had better be shortened. It is good, I think, to dash "in medias res," and work in later any descriptions of country, or any historical details which may be necessary. Murray likes lots of wood-cuts—give some by all means of ants. The public appreciate monkeys—our poor cousins. What sexual differences are there in monkeys? Have you kept them tame? if so, about their expression. I fear that you will hardly read my vile hand-writing, but I cannot without killing trouble write better.

You shall have my candid opinion on your MS., but remember it is hard to judge from MS., one reads slowly, and heavy parts seem much heavier. A first-rate judge thought my Journal very poor; now that it is in print, I happen to know, he likes it. I am sure you will understand why I am so egotistical.

I was a *little* disappointed in Wallace's book * on the Amazon; hardly facts enough. On other hand, in Gosse's book † there is not reasoning enough to my taste. Heaven knows whether you will care to read all this scribbling. . . .

I am glad you had a pleasant day with Hooker,‡ he is an admirably good man in every sense.

[The following extract from a letter to Mr. Bates on the same subject is interesting as giving an idea of the plan followed by my father in writing his 'Naturalist's Voyage:'

"As an old hackneyed author, let me give you a bit of advice, viz. to strike out every word which is not quite necessary to the current subject, and which could not interest a stranger. I constantly asked myself, Would a stranger care for this? and struck out or left in accordingly. I think too much pains cannot be taken in making the style transparently clear and throwing eloquence to the dogs."

Mr. Bates's book, 'The Naturalist in the Amazons,' was published in 1863, but the following letter may be given here rather than in its due chronological position :]

* 'Travels on the Amazon and Rio Negro,' 1853.
† Probably the 'Naturalist's Sojourn in Jamaica,' 1851.
‡ In a letter to Sir J. D. Hooker (Dec. 1861), my father wrote : " I am very glad to hear that you like Bates. I have seldom in my life been more struck with a man's power of mind."

C. Darwin to H. W. Bates.

Down, April 18, 1863.

DEAR BATES,—I have finished vol. i. My criticisms may be condensed into a single sentence, namely, that it is the best work of Natural History Travels ever published in England. Your style seems to me admirable. Nothing can be better than the discussion on the struggle for existence, and nothing better than the description of the Forest scenery.* It is a grand book, and whether or not it sells quickly, it will last. You have spoken out boldly on Species ; and boldness on the subject seems to get rarer and rarer. How beautifully illustrated it is. The cut on the back is most tasteful. I heartily congratulate you on its publication.

The *Athenæum* † was rather cold, as it always is, and insolent in the highest degree about your leading facts. Have you seen the *Reader?* I can send it to you if you have not seen it. . . .

C. Darwin to Asa Gray.

Down, Dec. 11 [1861].

MY DEAR GRAY,—Many and cordial thanks for your two last most valuable notes. What a thing it is that when you receive this we may be at war, and we two be bound, as good patriots, to hate each other, though I shall find this hating you very hard work. How curious it is to see two countries, just like two angry and silly men, taking so opposite a view of the same transaction ! I fear there is no shadow of doubt we shall fight, if the two Southern rogues are not given

---

* In a letter to Lyell my father wrote : " He [*i.e.* Mr. Bates] is second only to Humboldt in describing a tropical forest."

† " I have read the first volume of Bates's Book ; it is capital, and I think the best Natural History Travels ever published in England. He is bold about Species, &c., and the *Athenæum* coolly says 'he bends his facts' for this purpose."—(From a letter to Sir J. D. Hooker.)

up.*   And what a wretched thing it will be if we fight on the
side of slavery.   No doubt it will be said that we fight to get
cotton ; but I fully believe that this has not entered into the
motive in the least.   Well, thank Heaven, we private individuals have nothing to do with so awful a responsibility.
Again, how curious it is that you seem to think that you can
conquer the South ; and I never meet a soul, even those who
would most wish it, who thinks it possible—that is, to conquer
and retain it.   I do not suppose the mass of people in your
country will believe it, but I feel sure if we do go to war it
will be with the utmost reluctance by all classes, Ministers of
Government and all.   Time will show, and it is no use writing
or thinking about it.   I called the other day on Dr. Boott,
and was pleased to find him pretty well and cheerful.   I see,
by the way, he takes quite an English opinion of American
affairs, though an American in heart.†   Buckle might write
a chapter on opinion being entirely dependent on longitude !
    . . . With respect to Design, I feel more inclined to show
a white flag than to fire my usual long-range shot.   I like to
try and ask you a puzzling question, but when you return the
compliment I have great doubts whether it is a fair way of
arguing.   If anything is designed, certainly man must be :
one's "inner consciousness" (though a false guide) tells one
so ; yet I cannot admit that man's rudimentary mammæ . . .
were designed.   If I was to say I believed this, I should
believe it in the same incredible manner as the orthodox
believe the Trinity in Unity.   You say that you are in a
haze ; I am in thick mud ; the orthodox would say in fetid,
abominable mud ; yet I cannot keep out of the question.
My dear Gray, I have written a deal of nonsense.

Yours most cordially,

C. DARWIN.

---

* The Confederate Commis-
sioners Slidell and Mason were
forcibly removed from the *Trent,*
a West India mail steamer, on

Nov. 8, 1861. The news that the
U.S. agreed to release them reached
England on Jan. 8, 1862.

    † Dr. Boott was born in the U.S.

1862.

[Owing to the illness from scarlet fever of one of his boys, he took a house at Bournemouth in the autumn.   He wrote to Dr. Gray from Southampton (Aug. 21, 1862) :—

" We are a wretched family, and ought to be exterminated. We slept here to rest our poor boy on his journey to Bourne-mouth, and my poor dear wife sickened with scarlet fever, and has had it pretty sharply, but is recovering well.   There is no end of trouble in this weary world.   I shall not feel safe till we are all at home together, and when that will be I know not.   But it is foolish complaining."

Dr. Gray used to send postage stamps to the scarlet fever patient; with regard to this good-natured deed my father wrote—

" I must just recur to stamps ; my little man has calculated that he will now have 6 stamps which no other boy in the school has.   Here is a triumph.   Your last letter was plaistered with many coloured stamps, and he long surveyed. the envelope in bed with much quiet satisfaction."

The greater number of the letters of 1862 deal with the Orchid work, but the wave of conversion to Evolution was. still spreading, and reviews and letters bearing on the subject still came in numbers.   As an example of the odd letters he received may be mentioned one which arrived in January of this year " from a German homœopathic doctor, an ardent admirer of the ' Origin.'   Had himself published nearly the same sort of book, but goes much deeper.   Explains. the origin of plants and animals on the principles of ho-mœopathy or by the law of spirality.   Book fell dead in Germany.   Therefore would I translate it and publish it in. England."]

C. Darwin to T. H. Huxley.

Down, [Jan. ?] 14 [1862].

MY DEAR HUXLEY,—I am heartily glad of your success in the North,* and thank you for your note and slip. By Jove you have attacked Bigotry in its stronghold. I thought you would have been mobbed. I am so glad that you will publish your Lectures. You seem to have kept a due medium between extreme boldness and caution. I am heartily glad that all went off so well. I hope Mrs. Huxley is pretty well. . . . . I must say one word on the Hybrid question. No doubt you are right that here is a great hiatus in the argument; yet I think you overrate it—you never allude to the excellent evidence of *varieties* of Verbascum and Nicotiana being partially sterile together. It is curious to me to read (as I have to-day) the greatest crossing *Gardener* utterly pooh-poohing the distinction which *Botanists* make on this head, and insisting how frequently crossed *varieties* produce sterile offspring. Do oblige me by reading the latter half of my Primula paper in the 'Linn. Journal,' for it leads me to suspect that sterility will hereafter have to be largely viewed as an acquired or *selected* character—a view which I wish I had had facts to maintain in the 'Origin.' †. . . .

C. Darwin to J. D. Hooker.

Down, Jan. 25 [1862].

MY DEAR HOOKER,—Many thanks for your last Sunday's letter, which was one of the pleasantest I ever received in my life. We are all pretty well redivivus, and I am at work again. I thought it best to make a clean breast to Asa

* This refers to two of Mr. Huxley's lectures, given before the Philosophical Institution of Edinburgh in 1862. The substance of them is given in 'Man's Place in Nature.'

† The view here given will be discussed in the chapter on heterostyled plants.

Gray ; and told him that the Boston dinner, &c. &c., had quite turned my stomach, that I almost thought it would be good for the peace of the world if the United States were split up ; on the other hand, I said that I groaned to think of the slave-holders being triumphant, and that the difficulties of making a line of separation were fearful.   I wonder what he will say. . . . . Your notion of the Aristocrat being ken-speckle, and the best men of a good lot being thus easily selected is new to me, and striking.   The 'Origin' having made you in fact a jolly old Tory, made us all laugh heartily.   I have sometimes speculated on this subject ; primogeniture* is dreadfully opposed to selection ; suppose the first-born bull was necessarily made by each farmer the begetter of his stock !   On the other hand, as you say, ablest men are con-tinually raised to the peerage, and get crossed with the older Lord-breeds, and the Lords continually select the most beautiful and charming women out of the lower ranks ; so that a good deal of indirect selection improves the Lords. Certainly I agree with you the present American row has a very Torifying influence on us all.   I am very glad to hear you are beginning to print the ' Genera ;' it is a wonderful satisfaction to be thus brought to bed, indeed it is one's chief satisfaction, I think, though one knows that another bantling will soon be developing. . . .

C. Darwin to Maxwell Masters.†

Down, Feb. 26 [1862].

MY DEAR SIR,—I am much obliged to you for sending me

---

* My father had a strong feeling as to the injustice of primogeniture, and in a similar spirit was often indignant over the unfair wills that appear from time to time.   He would declare energetically that if he were law-giver no will should be valid that was not published in the testator's lifetime ; and this he maintained would prevent much of the monstrous injustice and mean-ness apparent in so many wills.

† Dr. Masters is a well-known vegetable teratologist, and has been for many years the editor of the *Gardeners' Chronicle.*

your article,* which I have just read with much interest. The History, and a good deal besides, was quite new to me. It seems to me capitally done, and so clearly written. You really ought to write your larger work. You speak too generously of my book ; but I must confess that you have pleased me not a little ; for no one, as far as I know, has ever remarked on what I say on classification,—a part, which when I wrote it, pleased me. With many thanks to you for sending me your article, pray believe me,

My dear Sir, yours sincerely,

C. DARWIN.

[In the spring of this year (1862) my father read the second volume of Buckle's 'History of Civilization.' The following strongly expressed opinion about it may be worth quoting :—

"Have you read Buckle's second volume ? it has interested me greatly ; I do not care whether his views are right or wrong, but I should think they contained much truth. There is a noble love of advancement and truth throughout ; and to my taste he is the very best writer of the English language that ever lived, let the other be who he may."]

C. Darwin to Asa Gray.

Down, March 15 [1862].

MY DEAR GRAY,—Thanks for the newspapers (though they did contain digs at England), and for your note of Feb. 18th. It is really almost a pleasure to receive stabs from so smooth, polished and sharp a dagger as your pen. I heartily wish I could sympathise more fully with you, instead of merely hating the South. We cannot enter into your feelings ; if Scotland were to rebel, I presume we should be very wrath, but I do not think we should care a penny what other nations

---

* A paper on " Vegetable Mor-   'British and Foreign Medico-Chi-
phology," by Dr. Masters, in the   rurgical Review ' for 1862.

thought. The millennium must come before nations love each other ; but try and do not hate me. Think of me, if you will as a poor blinded fool. I fear the dreadful state of affairs must dull your interest in Science. . . . .

I believe that your pamphlet has done my book *great* good ; and I thank you from my heart for myself ; and believing that the views are in large part true, I must think that you have done natural science a good turn. Natural Selection seems to be making a little progress in England and on the Continent ; a new German edition is called for, and a French * one has just appeared. One of the best men, though at present unknown, who has taken up these views, is Mr. Bates ; pray read his ' Travels in Amazonia,' when they appear ; they will be very good, judging from MS. of the first two chapters.

. . . . Again I say, do not hate me.

Ever yours most truly,

C. DARWIN.

C. Darwin to C. Lyell.

1 Carlton Terrace, Southampton,†
Aug. 22 [1862].

. . . . I heartily hope that you ‡ will be out in October. . . . . You say that the Bishop and Owen will be down on you ; the latter hardly can, for I was assured that Owen in his Lectures this spring advanced as a new idea that

---

* In June, 1862, my father wrote to Dr. Gray : " I received, 2 or 3 days ago, a French translation of the ' Origin,' by a Madlle. Royer, who must be one of the cleverest and oddest women in Europe : is an ardent Deist, and hates Christianity, and declares that natural selection and the struggle for life will explain all morality, nature of man, politics, &c. &c.! She makes some very curious and good hits, and says she shall publish a book on these subjects." Madlle. Royer added foot-notes to her translation, and in many places where the author expresses great doubt, she explains the difficulty, or points out that no real difficulty exists.

† The house of his son William.

‡ *I.e.* ' The Antiquity of Man.'

wingless birds had lost their wings by disuse, also that magpies stole spoons, &c., from a *remnant* of some instinct like that of the Bower-Bird, which ornaments its playing-passage with pretty feathers. Indeed, I am told that he hinted plainly that all birds are descended from one . . . .

Your P.S. touches on, as it seems to me, very difficult points. I am glad to see [that] in the 'Origin,' I only say that the naturalists generally consider that low organisms vary more than high ; and this I think certainly is the general opinion. I put the statement this way to show that I considered it only an opinion probably true. I must own that I do not at all trust even Hooker's contrary opinion, as I feel pretty sure that he has not tabulated any result. I have some materials at home, I think I attempted to make this point out, but cannot remember the result.

Mere variability, though the necessary foundation of all modifications, I believe to be almost always present, enough to allow of any amount of selected change ; so that it does not seem to me at all incompatible that a group which at any one period (or during all successive periods) varies less, should in the long course of time have undergone more modification than a group which is generally more variable.

Placental animals, e.g. might be at each period less variable than Marsupials, and nevertheless have undergone more *differentiation* and development than marsupials, owing to some advantage, probably brain development.

I am surprised, but do not pretend to form an opinion at Hooker's statement that higher species, genera, &c., are best limited. It seems to me a bold statement.

Looking to the 'Origin,' I see that I state that the productions of the land seem to change quicker than those of the sea (Chapter X., p. 339, 3rd edition), and I add there is some reason to believe that organisms considered high in the scale change quicker than those that are low. I remember writing these sentences after much deliberation. . . . . I remember well feeling much hesitation about putting in even the guarded sentences which I did. My doubts, I remember, related to the rate of change of the Radiata in the Secondary formation, and of the Foraminifera in the oldest Tertiary beds. . . . .

Good night,

C. DARWIN.

C. Darwin to C. Lyell.

Down, Oct. 1 [1862].

. . . . I found here * a short and very kind note of Falconer, with some pages of his 'Elephant Memoir,' which will be published, in which he treats admirably on long persistence of type. I thought he was going to make a good and crushing attack on me, but, to my great satisfaction, he ends by pointing out a loophole, and adds,† "with him I have no faith that the mammoth and other extinct elephants made their appearance suddenly. . . . . The most rational view seems to be that they are the modified descendants of earlier progenitors, &c." This is capital. There will not be soon one good palæontologist who believes in immutability. Falconer does not allow for the Proboscidean group being a failing one, and therefore not likely to be giving off new races.

He adds that he does not think Natural Selection suffices. I do not quite see the force of his argument, and he apparently overlooks that I say over and over again that Natural Selection can do nothing without variability, and that variability is subject to the most complex fixed laws. . . . .

[In his letters to Sir J. D. Hooker, about the end of this

---

* On his return from Bournemouth.

† Falconer, "On the American Fossil Elephant," in the 'Nat. Hist. Review,' 1863, p. 81. The words preceding those cited by my father make the meaning of his quotation clearer. The passage begins as follows : "The inferences which I draw from these facts are not opposed to one of the leading propositions of Darwin's theory. With him," &c. &c.

year, are occasional notes on the progress of the 'Variation of Animals and Plants.' Thus on November 24th he wrote: "I hardly know why I am a little sorry, but my present work is leading me to believe rather more direct in the action of physical conditions. I presume I regret it, because it lessens the glory of Natural Selection, and is so confoundedly doubtful. Perhaps I shall change again when I get all my facts under one point of view, and a pretty hard job this will be."

Again, on December 22nd, "To-day I have begun to think of arranging my concluding chapters on Inheritance, Reversion, Selection, and such things, and am fairly paralysed how to begin and how to end, and what to do, with my huge piles of materials."]

C. Darwin to Asa Gray.

Down, Nov. 6 [1862].

MY DEAR GRAY,—When your note of October 4th and 13th (chiefly about Max Müller) arrived, I was nearly at the end of the same book,* and had intended recommending you to read it. I quite agree that it is extremely interesting, but the latter part about the *first* origin of language much the least satisfactory. It is a marvellous problem. . . . . [There are] covert sneers at me, which he seems to get the better of towards the close of the book. I cannot quite see how it will forward "my cause," as you call it; but I can see how any one with literary talent (I do not feel up to it) could make great use of the subject in illustration.† What pretty metaphors you would make from it! I wish some one would

* 'Lectures on the Science of Language,' 1st edit. 1861.

† Language was treated in the manner here indicated by Sir C. Lyell in the 'Antiquity of Man.'

Also by Prof. Schleicher, whose pamphlet was fully noticed in the *Reader*, Feb. 27, 1864 (as I learn from one of Prof. Huxley's 'Lay Sermons').

keep a lot of the most noisy monkeys, half free, and study their means of communication!

A book has just appeared here which will, I suppose, make a noise, by Bishop Colenso,* who, judging from extracts, smashes most of the Old Testament. Talking of books, I am in the middle of one which pleases me, though it is very innocent food, viz. Miss Cooper's 'Journal of a Naturalist.' Who is she? She seems a very clever woman, and gives a capital account of the battle between *our* and *your* weeds. Does it not hurt your Yankee pride that we thrash you so confoundedly? I am sure Mrs. Gray will stick up for your own weeds. Ask her whether they are not more honest, downright good sort of weeds. The book gives an extremely pretty picture of one of your villages ; but I see your autumn, though so much more gorgeous than ours, comes on sooner, and that is one comfort. . . . .

C. Darwin to H. W. Bates.

Down, Nov. 20, [1862].

DEAR BATES,—I have just finished, after several reads, your paper.† In my opinion it is one of the most remarkable and

* 'The Pentateuch and Book of Joshua critically examined,' six parts, 1862–71.

† This refers to Mr. Bates's paper, "Contributions to an Insect Fauna of the Amazons Valley" ('Linn. Soc. Trans.' xxiii., 1862), in which the now familiar subject of mimicry was founded. My father wrote a short review of it in the 'Natural History Review,' 1863, p. 219, parts of which occur almost verbatim in the later editions of the 'Origin of Species.' A striking passage occurs showing the difficulties of the case from a creationist's point of view :—

" By what means, it may be asked, have so many butterflies of the Amazonian region acquired their deceptive dress? Most naturalists will answer that they were thus clothed from the hour of their creation—an answer which will generally be so far triumphant that it can be met only by long-drawn arguments ; but it is made at the expense of putting an effectual bar to all further inquiry. In this particular case, moreover, the creationist will meet with special difficulties ; for many of the mimicking forms of *Leptalis* can be shown by a graduated series to be merely

admirable papers I ever read in my life. The mimetic cases are truly marvellous, and you connect excellently a host of analogous facts. The illustrations are beautiful, and seem very well chosen; but it would have saved the reader not a little trouble, if the name of each had been engraved below each separate figure. No doubt this would have put the engraver into fits, as it would have destroyed the beauty of the plate. I am not at all surprised at such a paper having consumed much time. I am rejoiced that I passed over the whole subject in the 'Origin,' for I should have made a precious mess of it. You have most clearly stated and solved a wonderful problem. No doubt with most people this will be the cream of the paper; but I am not sure that all your facts and reasonings on variation, and on the segregation of complete and semi-complete species, is not really more, or at least as valuable, a part. I never conceived the process nearly so clearly before; one feels present at the creation of new forms. I wish, however, you had enlarged a little more on the pairing of similar varieties; a rather more numerous body of facts seems here wanted. Then, again, what a host of curious miscellaneous observations there are—as on related

varieties of one species; other mimickers are undoubtedly distinct species, or even distinct genera. So again, some of the mimicked forms can be shown to be merely varieties; but the greater number must be ranked as distinct species. Hence the creationist will have to admit that some of these forms have become imitators, by means of the laws of variation, whilst others he must look at as separately created under their present guise; he will further have to admit that some have been created in imitation of forms not themselves created as we now see them, but due to the laws of variation! Prof. Agassiz, indeed, would think nothing of this difficulty; for he believes that not only each species and each variety, but that groups of individuals, though identically the same, when inhabiting distinct countries, have been all separately created in due proportional numbers to the wants of each land. Not many naturalists will be content thus to believe that varieties and individuals have been turned out all ready made, almost as a manufacturer turns out toys according to the temporary demand of the market."

sexual and individual variability : these will some day, if I live, be a treasure to me.

.With respect to mimetic resemblance being so common with insects, do you not think it may be connected with their small size ; they cannot defend themselves ; they cannot escape by flight, at least, from birds, therefore they escape by trickery and deception ?

I have one serious criticism to make, and that is about the title of the paper ; I cannot but think that you ought to have called prominent attention in it to the mimetic resemblances. Your paper is too good to be largely appreciated by the mob of naturalists without souls ; but, rely on it, that it will have *lasting* value, and I cordially congratulate you on your first great work. You will find, I should think, that Wallace will fully appreciate it. How gets on your book ? Keep your spirits up. A book is no light labour. I have been better lately, and working hard, but my health is very indifferent. How is your health ? Believe me, dear Bates,

Yours very sincerely,
C. DARWIN.

END OF VOL. II.

Printed in the United States
By Bookmasters